A PLAYER'S

GUIDE TO THE

POST-TRUTH

CONDITION

KEY ISSUES IN MODERN SOCIOLOGY

The Sociology programme takes a fresh and challenging sociological look at the interactions between politics, society, history and culture. Titles transcend traditional disciplinary boundaries. This programme includes a variety of book series.

Key Issues in Modern Sociology publishes scholarly texts by leading social theorists that give an accessible exposition of the major structural changes in modern societies. These volumes address an academic audience through their relevance and scholarly quality, and connect sociological thought to public issues. The series covers both substantive and theoretical topics, as well as addressing the works of major modern sociologists. The series emphasis is on modern developments in sociology with relevance to contemporary issues such as globalization, warfare, citizenship, human rights, environmental crises, demographic change, religion, postsecularism and civil conflict.

A Player's

GUIDE TO THE POST-TRUTH CONDITION

The Name of the Game

STEVE FULLER

ANTHEM PRESS

Anthem Press
An imprint of Wimbledon Publishing Company
www.anthempress.com

This edition first published in UK and USA 2020
by ANTHEM PRESS
75–76 Blackfriars Road, London SE1 8HA, UK
or PO Box 9779, London SW19 7ZG, UK
and
244 Madison Ave #116, New York, NY 10016, USA

British Library Cataloguing-in-Publication Data
A catalogue record for this book is available from the British Library.

ISBN-13: 978-1-78527-603-3 (Hbk)
ISBN-10: 1-78527-603-4 (Hbk)
ISBN-13: 978-1-78527-606-4 (Pbk)
ISBN-10: 1-78527-606-9 (Pbk)

This title is also available as an e-book.

CONTENTS

A WORD TO THE READER

This book is designed to do what its title says, namely, to provide a guide to the post-truth condition for those who wish to feel at home and thrive in it – rather than simply avoid or attack it. It consists of a series of short chapters that are best read in the order presented but may also be read in a different order or simply in parts – as most books are normally read. The book ranges widely across philosophy, theology, science, politics, economics, psychology, and the arts – but hopefully in a way that allows readers to find their bearings, given the opportunities presented by the internet to follow up whatever might interest them in the text. Underlying this breadth of scope is a fundamental scepticism with 'business as usual' in the production and evaluation of knowledge claims. To be sure, the reader will see that post-truth extends many of the themes already found in what passes for 'postmodernism'. However, at a deeper level, and in light of the ongoing COVID-19 pandemic, the post-truth condition invites us to discover in a new key what it has always meant to be 'modern'.

ACKNOWLEDGEMENTS

In writing this book, I kept in mind three quite different people who had already embodied the post-truth condition in their beliefs and action early in my academic career nearly forty years ago: Deirdre McCloskey, Charles Arthur Willard and Steve Woolgar.

During this book's composition, the following people offered insight, provocation and support of various sorts: Aleksandra Łukaszewicz Alcaraz, Thomas Basbøll, Anya Bernstein, Kean Birch, Yael Brahms, Sindi Breshani, Mattia Gallotti, Jane Gilbert, Zhengdong Hu, Petar Jandric, Ian Jarvie, David Johnson, Paul Jump, Ilya Kasavin, Josephine Lethbridge, David Levy, Veronika Lipinska, Luke Robert Mason, Linsey McGoey, Alfred Nordmann, Nathan Oseroff, Lea Peersman Pujol, Ljiljana Radenović, Sheldon Richmond, Sharon Rider, Raphael Sassower, Nico Stehr, Adam Tait, Gareth Thompson, Steven Umbrello and Cong Wang.

Finally, this book is dedicated to my mother, who died in New York City during the COVID-19 pandemic – but not of it. But that's a story for another day …

INTRODUCTION: HOW TO LEARN TO STOP WORRYING AND LOVE THE POST-TRUTH CONDITION

The post-truth condition means many things to many people. Simply put and without prejudice, *in the post-truth condition, what matters is not whether something is true or false but how the matter is decided.* As we shall see in the following pages, there is a lot more to it than that. But that is the core meaning – and it is already enough to strike fear in the hearts of many. Opponents of the post-truth condition identify it with a credibility crisis in knowledge-based institutions that stems from some evil force intent on leading a gullible public to undo all the careful work that over the past five centuries has made ours an increasingly rational world in which people can operate freely to mutual benefit. These malign manipulators range from alternative newsfeeds such as Breitbart to big data firms such as Cambridge Analytica to the vague but looming presence of 'Russian hackers'. I take a very different approach to the post-truth condition.

I welcome the post-truth condition as nothing more – and nothing less – than the logical next stage of the very same project of rationalisation that post-truth's opponents claim to uphold. It should not be feared but embraced as a sign of this project's genuine democratisation. Authority is finally being devolved from a vanguard class of 'experts' with a monopoly of moral and political force to some yet-to-be-defined organisation of independent self-legislating individuals. In the coming years we should expect that such modern 'establishment' institutions as the 'state' and the 'university' will be subject to the same shakedown that the 'church' has periodically undergone since the early modern period in parts of the world touched by Christianity. To be sure, the exact sort of 'organisation' that governs the post-truth condition is very much up for grabs, and may always be, which explains the subtitle of this book: *The Name of the Game.* The ultimate prize in the post-truth condition is to name the game you play, even if you turn out to be the loser. Put as a point of strategy, history can be written by the losers, if they manage to make the winners feel guilty about having won, thereby handicapping them in any subsequent game. Readers of Nietzsche will understand exactly what I mean.

What is most evident about the post-truth condition is that the difference between those who know and those who don't has been reduced – but not quite as you might think. To be sure, we live in a time of unprecedented levels of literacy, schooling and access of information, notwithstanding what opponents of the post-truth condition sometimes seem to assume. Thus, lay people can increasingly catch up with what the experts know to the point of confidently challenging their judgement, even if they are rebuffed in the end. And no doubt this dynamic goes to the democratic heart of the post-truth condition. But the sort of 'reduction' I really have in mind is about the playing field of uncertainty becoming levelled.

In the past, experts have exercised a form of 'cognitive authoritarianism' (Fuller 1988: ch. 12). They have enjoyed a monopoly licence to turn uncertainty to their advantage by dictating the name of the game that everyone else should be playing. This 'monopoly licence' has normally taken the form of state-certified academic credentials that grant experts the right to heroically simplify reality in the name of policy. Its signature contemporary practice is 'modelling', a mathematically inspired and technologically enhanced descendant of the esoteric art of the 'microcosm'. Both involve a sophisticated way of extending from what one knows to what one doesn't know in order to create a template for reality as a whole. In the European Renaissance this art was typically depicted in Platonic terms as a rigorous way of recalling a collective past but nowadays it is discussed more 'naturalistically' as the search for underlying causes. And so the magicians yielded to the scientists (Yates 1966). In any case, in the post-truth condition, that monopoly over modelling is broken, resulting in a free market with multiple competitors which effectively democratises control over uncertainty.

It is no longer simply that the world itself is uncertain but who is best positioned to manage that uncertainty is also uncertain. It is, as the logicians would say, this 'second order' shrinkage in the epistemic distance between experts and lay people that ultimately defines the post-truth condition. This helps to explain the increasingly heavy-handed ways in which experts have tried to exert their authority in the public sphere, very much like an autocrat who needs to regularly remind subjects of his power, lest it slip away. As Machiavelli and others have realised, this is a sign of weakness not strength, since the sheer survival of nonconformists forces the autocrat to redouble his efforts to reassert his authority. And once the autocrat takes the bait and redoubles his efforts, he will effectively recognise his opponents as equals, as well as expend his own energies in a self-defeating cause. Typically the autocrat will be forced to settle for a world in which he perhaps retains dominance but not hegemony. In this spirit, I believe that the post-truth condition will leave the university – and the scientific establishment as a whole – in the position of the Roman Catholic Church after the Protestant Reformation vis-à-vis Christendom.

This prospect is at once exciting and scary. For the full measure of what I mean, consider the following potted history of human institutions which sociologists have been telling since the late nineteenth century. In the beginning, institutions were reproduced on a hereditary basis, as social lineage closely tracked biological lineage. The path to modernity started to be paved once the lines of social and biological descent were disaggregated. It began with the clergy and the military, and gradually spread across the productive sectors of society. By the end of the eighteenth century, the presumed basis for institutional reproduction had shifted away from hereditary entitlement to examination and election. Periodic performance checks increasingly replaced the need to wait for an incumbent to die or resign. At that point, institutions had effectively become disaggregated as 'corporate' entities from the individuals who happen to carry out their functions at a given time. This change corresponded to a shift in the source of institutional legitimacy from some original moment in history to a 'charter' or 'constitution' that functions as a sustaining generative programme. It is not so different from how we think about a computer algorithm today, in which theoretically the same programme can run on many different machines indefinitely, without being dependent on any particular set of individuals.

The post-truth condition marks a third disaggregation in this potted history. After all, a state whose leaders are constitutionally subject to regular elections is just as monopolistic as one whose leaders are subject to hereditary succession. Similarly, if the most influential theorist of science in the second half of the twentieth century is to be believed, the cognitive authority enjoyed by the dominant research paradigm in a field of empirical inquiry is no less monolithic than that enjoyed by an established church (Kuhn 1970). Against this backdrop, the post-truth condition amounts to 'epistemic trust-busting', resulting in a free market whereby multiple constitutions, lineages, paradigms and churches compete to name the game in their respective field of play (Fuller 2018: 48). One person's sense of the established order thus becomes another's conspiracy against the public interest. But more important than the sheer plurality and conflict of perspectives unleashed by this situation is the mental agility required to thrive in it. A successful player needs to see things from the standpoint of one's opponent, and if possible turn that to one's own advantage. If one is a chess player who encounters someone who plays checkers, then one's strategy must be to achieve an outcome in checkers comparable to the outcome one would wish to achieve in chess.

This characterisation of the post-truth condition will undoubtedly strike many philosophically literate people as no more than a sophisticated version of *relativism*. However, this would be to seriously underestimate the intellectual stakes. Relativism may be acceptable or abhorrent but it is philosophically straightforward. As its name suggests, 'relativism' presupposes a jurisdictional approach to claims about knowledge, morals and so forth. 'When in Rome, do as the Romans do', as Bishop Ambrose advised the young St Augustine. That's relativism in a nutshell, and hence its appeal to anthropologists who approach a native tribe as absolute outsiders. Indeed, relativism makes most sense if you are an outsider to all possible jurisdictions. Suppose you're looking at a map trying to figure out how to blend in with the inhabitants in various locations. It is the world view of tourists and chameleons. In contrast, the post-truth condition assumes that the jurisdictional boundaries are themselves up for grabs, which means that who counts as an 'insider' and who as an 'outsider' is never engraved in stone, let alone fixed in constitutional print. You're always in the thick of it, but there is always room to manoeuvre, which means that everyone is kept on their toes. The jurisdiction in which one is operating – the rules of the game – is constructed as one goes along. It is here that the post-truth condition's 'second-order' character comes to the fore.

If some readers feel the ground slipping beneath their feet, it is only because they are experiencing what it feels like to 'level the playing field'. The post-truth condition's sense of reality is intimately tied to its sense of fairness. *To be real is to be redeployable*. This motto may sound stark but it is quite familiar. Suppose we stick to 'fact' in its colloquial sense as a building block of some common understanding of reality, regardless of how the fact was discovered or constructed. If you were required to hold too many beliefs of a certain sort before being allowed to use a fact in argument, you would probably doubt its status as a 'fact' in this sense. The post-truth condition accepts this colloquial understanding of 'fact' which implies that it is a resource that can be deployed by all parties in multiple capacities and in multiple arguments. What is good for the goose is good for the gander: a fact for you is a fact for me. The implied premise of US Senator Daniel Patrick Moynihan's memorable quip, 'You are entitled to your own beliefs but not your own facts', is that facts are not the sort of thing that can be 'owned'. The post-truth condition simply underscores that facts can't be owned *even by the experts*. While you and I may disagree over how to deploy a fact in a particular context, not least because we may be struggling over how to define our common field of play, nevertheless the struggle is fair just as long as each of us has an equal opportunity to turn the facts to our respective advantage. A good way to envisage the situation is in terms of the famed test figures in Gestalt psychology whereby, say, the same stylised drawing (i.e. 'facts') can be interpreted as a duck or a rabbit, depending on the interpretative cues that the experimenter provides the subject.

This characterisation of the post-truth condition should be familiar from a variety of settings, many of which are invoked in this book. But two sorts of settings are worth highlighting at the outset. The first is the 'one person, one vote' principle that is common to modern democratic elections, utilitarian moral calculations and survey-based research. The second consists in the various methods that philosophers and scientists have developed from Francis Bacon onwards to ensure 'fair testing' in experimental and clinical settings. In both settings, fairness is engineered by manipulating uncertainty to some collective advantage, be it to the body politic or a body of knowledge. Moreover, two sorts of uncertainty are manipulated at once. On the one hand, there is the *spontaneous uncertainty* that is inherent in our relationship to the future, simply because we have yet to experience it. On the other hand, there is the *manufactured uncertainty* that comes from strategically suspending our personal histories. This typically involves minimising or even eliminating whatever cumulative advantage one party might already enjoy over another by virtue of, say, superior education, inherited wealth, prior track record or simply sheer longevity. Logicians like to say that a fair argument distributes the burden of proof equally among the contesting parties. It also captures the sense of 'fairness' endemic to the post-truth condition – as well as of the most influential theory of justice of the past half-century (Rawls 1971).

As I complete this book, 'Nature' in the form of the COVID-19 pandemic is performing its own version of 'levelling the playing field', perhaps more effectively than humans in their various efforts at mass democratisation and wealth redistribution down through the ages (Scheidel 2017). Who would have thought at the start of 2020 that a new virus might be poised to deliver on all the unfulfilled promises of both socialists and ecologists, who themselves have only intermittently joined in common cause? Of course, the actual delivery on those promises will depend on the ultimate length and shape of the pandemic. But clearly, the longer it lasts and its impact more pervasive, the post-truth condition will come increasingly to the fore, as everyone is forced to think in second-order terms about how the rules of the game need to be changed to their advantage.

But even without the virus, the post-truth condition was already reshaping the scientific establishment along the lines that I had begun to suggest in *Post-Truth: Knowledge as a Power Game* (Fuller 2018). A telling moment came in 2019 when the British Academy awarded its two main medals to Naomi Oreskes and Ben Goldacre, whose work 'brought academic rigour into public debate on the major challenges of our day', according to the Academy's press release (British Academy 2019). As it turns out, while Oreskes and Goldacre certainly see themselves and are seen by others as defenders of science, they are also 'self-styled'. Think freelancers. Neither really belongs to the scientific establishment, though both are scientifically trained. More importantly, and in keeping with the post-truth condition, their respective self-positioning pulls the establishment in opposing directions. Oreskes clearly wants to align science with the sources of power, whereas Goldacre wants to use science to dissolve those power sources. In the Machiavellian terms introduced in Fuller (2018), Oreskes is an angry 'lion' and Goldacre a furtive 'fox'. This odd couple expose the scientific establishment in ways that should worry the true Machiavellian. Whereas Oreskes wants scientists to assert their epistemic privilege more openly, Goldacre wants to use science against many who already enjoy that privilege. Oreskes and Goldacre are, respectively, Jesuit and Protestant in science's current Post-Truth Reformation.

Oreskes is a US science historian best known for her full-throated defence of a 'scientific consensus' on a host of policy-relevant matters ranging from smoking to climate change (Oreskes and Conway 2011). She regularly urges scientists to close ranks to combat the combined forces of populist and corporate interests that she believes have commandeered democratic politics. Indeed, she has provocatively suggested that China is better placed than the West to tackle any impending climate catastrophe because its authoritarianism avoids the need to persuade a fickle and ignorant electorate before taking decisive action (Oreskes and Conway 2014). Whatever 'science' might mean for Oreskes, it is certainly not Karl Popper's (1946) 'open society'! In contrast, Goldacre made his reputation as the 'Bad Science' columnist for London's *Guardian* newspaper, where he deployed his first-class Oxford training in medicine and statistics to deconstruct articles published in a variety of venues, ranging from popular media to peer-reviewed scientific journals. Like a magician revealing the tricks of his trade, Goldacre carefully showed the likely places where authors were deceiving their readers, if not themselves, given the conclusions drawn from the data presented (Goldacre 2008). He has increasingly focused on the pharmaceutical industry, the primary funder of biomedical research worldwide, where the biases and errors found in its sponsored publications can be easily shown to have been motivated by 'special interests' (Goldacre 2012).

A good general measure of the academic establishment's reluctance – if not outright failure – to acknowledge the post-truth condition is that we academics continue to give students the false impression that knowledge of enduring value is best backed by academically credible sources, which can then be dutifully transferred through the appropriate channels to enlighten the masses, who in turn apply that knowledge to the betterment of society. Of course, the history of knowledge doesn't conform to this neat linear narrative, but we academics have convenient ways to deal with the 'outliers' – who may include the majority of actual knowledge producers. When the outliers are altogether irreconcilable to the academic narrative, they are portrayed as having interrupted and possibly diverted knowledge from its natural course. The more persistent that these outliers seem to be, the more malign their intent is made to appear. However, other outliers can be reconciled more easily. Their claims to knowledge emerged 'idiosyncratically' or 'by accident', but were eventually 'justified' by the appropriate academic authorities. Indeed, some of these outliers may be dignified as having been 'ahead of their time'.

However, what is typically absent from academic theorists of knowledge – be they philosophers or sociologists – is much serious discussion about certain knowledge claims being rejected simply due to the sheer difficulty of assimilating them without radically altering the dominant narrative of progress in knowledge. In this context, 'difficulty' is ambiguous between strictly logical and more openly political issues. This is understandable because to shift the truth value of a proposition is effectively to alter the authority of someone who upholds that proposition. It is arguably the first lesson of the post-truth condition. So it should come as no surprise that the most influential epistemologist of the second half of the twentieth century, Willard Quine, was clear that even a knowledge claim well evidenced on its own terms but whose acceptance would require a massive change to 'our' overall 'web of belief' should be treated as dubious until further notice (Quine and Ullian 1970). Here 'our' should be understood as a euphemism for those of us who have been academically domesticated or have a stake in the dominant academic knowledge game.

What follows in these pages can be read as turning the academic narrative about knowledge production on its head. Genuinely innovative knowledge has come mainly from *outside* the academy, which typically lags behind because of its centrality to the production and distribution of *credentials* which, in turn, involves the conversion of knowledge into the DNA that informs societal reproduction. Nevertheless, that centrality has enabled academia to co-opt with relative ease extramurally produced forms of knowledge through *education*, the core activity associated with the acquisition of credentials. Indeed, postmodernism's founding text, which was commissioned as a report on the state of higher education, is largely a meditation on this fact (Lyotard 1983). Genuine innovations in the modern era have been the products of 'military-industrial' knowledge interests, which universities subsequently captured (Fuller 2018: ch. 4). To be sure, universities have also improved, extended and made that knowledge more generally available, thereby ensuring that what might have remained intellectual property enters the public domain. While hardly a trivial achievement, it has been often made in the face of resistance from the academics themselves. As we shall see, the self-selecting and self-certifying character of academia's 'peer review' processes is a source of the problem. Nevertheless, as the changing political economy of academic funding and publishing has brought it much closer to the information ecology of social media, the playing field between academic and non-academic forms of knowledge is being levelled, the ultimate sign that we are truly in the midst of the post-truth condition.

1

POST-TRUTH BREAKS FREE OF REASON'S OWN SELF-IMPOSED CHAINS

The post-truth condition is stereotyped by its detractors as the subversion of reason by factors external to it, such as emotions, bias and lack of evidence. But for post-truth's devotees, reason is all too often subverted by its failure to be applied in a principled manner. We let extraneous beliefs curtail a line of thought that is open to a more radical exploration than we assume. The political psychologist Philip Tetlock (2003) coined the resonant phrase *taboo cognition* for those limits that we routinely – and artificially – impose on our own reasoning. A good example of a taboo cognition and its violation is the self-restraint that had been exercised by Christian theologians about the human character of Jesus – that is, until the late nineteenth century, when a very active debate erupted over the 'psychiatric' state of humanity's purported saviour, implying that he might have suffered from a mental disorder. Suddenly what had previously been taboo was discussed openly. The great humanitarian Albert Schweitzer devoted his 1911 medical doctorate to this topic, in which he famously – and perhaps decisively – defended Jesus's sanity.

What is interesting about this debate from a post-truth standpoint is that it took almost 2000 years for it to happen, even though Jesus's centrality to Christianity lies in his status as part-human and part-divine, which serves as an ever-present lifeline to humanity's salvation from its fallen state. While there had been much discussion about the weakness of Jesus's body prior to the late nineteenth century, there had been little if any discussion of his potential weakness of mind. Past sceptics had largely ignored the Jesus narrative rather than probe the limits of its credibility. They never took the

Bible seriously enough to test it. Indeed, 'criticism' came into usage as a contrast to 'scepticism' only in the late eighteenth century to capture such taboo-breaching probing. To be sure, Christianity survived the probes about Jesus's mental health, albeit by regrouping (aka 'modernising') over the twentieth century. While its default frame of reference had been assaulted and forced to adapt, it was not destroyed. However, not all frames of reference are so resilient, yet the post-truth condition is designed to test them all. Is there a generic strategy lurking here?

The ancient Greeks first hit upon the matter when pondering the *sorites paradox*, which logicians have popularised as the 'slippery slope argument'. Let's start with a simple example from logic textbooks and then shift register to the forbidden zone of theology. We seem to have a clear, perhaps even absolute sense of the difference between 'hairy' and 'bald', and we assume that many other value judgements – say, concerning someone's age – hang on that difference. Clearly a full head of hair makes one 'hairy'. But suppose just a single hair is removed: Isn't the person still 'hairy'? Of course. But then you keep removing hairs, one at a time. It is equally clear that at some point the person is no longer 'hairy' but 'bald' – but when? The lesson learned by post-truth devotees is that once a binary such as 'hairy' versus 'bald' is admitted to be a continuum that admits of overlapping degrees of 'hairiness' or 'baldness', then intuitions relating to the values carried by the distinction start to be lost. One realises that a hairy person might be old, while a bald one might be young.

Now, transfer this line of thought to Christian theology. Here we start by assuming a clear, perhaps even absolute difference between 'human' and 'divine'. And of course, many other value judgements hang on this difference, as 'divine' is regarded in all respects as superior to 'human'. However, Christians also believe that humans were not created this way; on the contrary, the Bible says that we were created 'in the image and likeness of God'. Of course, our species suffered a 'Fall from Grace' following the first human's partaking of the forbidden fruit, which resulted in all subsequent generations of humans being tainted by 'Original Sin'. St Augustine influentially characterised this fallen state as our being 'deprived' of God. In other words, Adam's transgression did not turn humanity into God's mortal enemy but simply made us distant from the deity. If the Bible is right that God's original physical manifestation was as light, then humanity was rendered a planet cast far from the source of that light, as God effectively gave us the cold shoulder and walked away. We lost his frame of reference. This explains not only Dante's famous depiction of Hell as a realm mired in ice but also the centrality of optics to the slow medieval march to modern science as a source of human emancipation and salvation (Crombie 1996; Harrison 2007).

All of this redolent Christian imagery should not obscure that the difference between the divine and the human is ultimately one of *degree*, not *kind*. For humanity to be 'fallen' is simply for us to be on the opposite end of a continuum from God. It opens the prospect of transitioning back to a divine state, in which the hybrid status of Jesus serves as a hopeful precedent and exemplar. But of course, going down this route ultimately undermines the value judgements that had been tied to a clear distinction between the human and the divine. For example, the very idea of a 'church' governed by officials whose 'holiness' sets them apart from the rest of humanity as a 'priesthood' by virtue of their presumed closeness to God starts to look suspect, especially once the church's secular foundations are subject to scrutiny. This in a nutshell is how the interrogation of ancient entitlements and current clerical practices in the Renaissance led to the Protestant Reformation, which effectively democratised Christendom, which in turn paved the way to modernisation, arguably culminating in today's 'transhumanism'. That path, which runs through the heart of the West's self-understanding, has been clearly taboo-shattering.

What unites both the rather mundane hairy/bald binary and the much more momentous divine/human binary is that a situation which at first seemed simply *ambiguous* comes to be seen as *vague* – and that technical shift in semantic status carries explosive consequences for important value distinctions that ride on the respective binaries. The relevant semantic shift is between finding it hard to decide whether someone is hairy/bald or divine/human (ambiguous) and finding it hard to decide whether anything is really at stake because the distinction may be meaningless (vague). If 'hairy' and 'bald' doesn't reliably track 'young' and 'old', or 'divine' and 'human' doesn't reliably track 'sacred' and 'profane', who cares? When this question starts to be asked, then the frame of reference that previously defined true and false, fact and fiction, becomes relaxed if not destroyed altogether. At that point, one enters the post-truth condition. To recall the start of our discussion, once not even Christian theologians could agree the basis on which to decide the divinity or humanity of Jesus, they started to evaluate Jesus by criteria on which they could potentially agree, such as his mental health. And following Albert Schweitzer, we might never know whether Jesus was divine or human, but at least we can agree he was sane – and that is enough.

Understanding the post-truth condition helps to explain why the outcry over so-called fake news is ultimately misdirected. Instead of focusing on how to tell the difference between 'real' and 'fake' news, we should be asking the prior question of whether the difference matters, especially if all 'news' contains at least a grain of truth and a degree of falseness. Facts are inherently vague unless we agree a frame of reference in which to evaluate them as 'evidence' – in the sense of something that matters to us all. Indeed, what is truly at stake in the various claims and counterclaims about 'fakeness' is not truth and veracity but *legitimacy* and *authority*. This long-standing impression has only been enhanced by the advent of social media, which allows for multiple competing news sources with access to mass audiences without the strong political and/or economic filters that had functioned as 'standards' when there were relatively few broadcasters. Thus, the burden is increasingly placed on individuals to decide what – and to what extent – to believe about some matter of interest to them. Which frame of reference should they adopt? The answer will turn on the exact nature of their interests in knowledge, their *reason* for knowing.

Here it is worth recalling that, philosophically speaking, 'reason' is more than simply the intellect. The intellect is the faculty that aims to reflect the nature of reality, which requires a studied receptiveness to the world as it is, and so contemplation has been seen as the defining exercise of the intellect. In contrast, 'reason' has been traditionally depicted as a much more active faculty, one associated with judgement and decision making. As a psychological formula: *Reason = Intellect + Will*. In effect, reason puts you in the driver's seat of reality. Instead of your being open to the world, the world is open to you. The cultivation of this assertive conception of reason was central to the training of the ideal rulers – the 'philosopher-kings' – in Plato's *Republic*. However, reason started to be seen as potentially within everyone's reach, with greater acceptance of St Augustine's particular interpretive spin on the Bible. God creates by literally dictating things into being (the divine *logos*), and as creatures in the 'image and likeness' of God we can do godlike things too by making reasoned choices based on first principles about the sort of world in which we would like to live. Of course, as we have seen, Original Sin places us at a strategic disadvantage but we may come to recover our own 'voice' as world-creators.

To be sure, this is the source of the distinctive 'Faustian' mentality of the West, whose obituary Oswald Spengler had already written in the aftermath of the First World War. It truly took hold in both science and politics during the eighteenth century-European Enlightenment. Over the past three hundred years it inspired an unprecedented mobilisation of human effort and natural resources to construct the much vaunted 'heaven on earth'. And of course, the trail of Enlightenment projects is littered with people who underestimated the difference between humans and gods. However, in the post-truth condition, rationality consists in gaming that difference. Here one might take a page from that self-styled maverick 'risk engineer', Nassim Nicholas Taleb (2012). For Taleb, the 'antifragile' investor is one who envisages the future as substantially different from the past but is uncertain exactly when and how that will come about. Taleb's preferred strategy is a kind of spread betting in which you anticipate many losses on the wilder bets, but you treat these as investments in learning from error. In short, the edge of reason lies in your capacity to make strategically designed mistakes in a frame of mind sufficiently open that you do not make them again.

2

POST-TRUTH IS ABOUT FINDING
A GAME ONE CAN WIN

It is misleading to think about 'post-truth' as entailing a *disregard* for truth. Even when Plato in the *Theaetetus* presented what philosophers generally still regard as the working definition of knowledge – 'justified true belief' – his emphasis was on the 'justified' rather than the 'true'. Plato's paradigm case was a trial lawyer who manages to get a jury to reach the correct judgement in a case, but for the wrong reasons. Plato seemed to think that this sort of thing happened frequently. And he may have been right. More to the point, he thought that it is something that should worry us. In contrast, the post-truth condition amounts to a demystification – if not an outright rejection – of Plato's worry. The 'post-truther' claims that fuss over how one justifies beliefs simply aims to bias, spin or otherwise restrict the course of inquiry. Thus, *contra* Plato, the lawyer is right to be primarily concerned with winning the case – that is, to establish the truth to his client's satisfaction (aka acquittal) under the conditions of play by whatever means he can.

To be sure, what the post-truther identifies as factors that 'bias, spin and otherwise restrict the course of inquiry', more circumspect philosophers would discuss in terms of whatever collateral 'values' and 'ends' might be served by acquiring knowledge of the truth. Unfortunately, that broader discussion – philosophers used to call it 'axiology' – can be obscured by academic talk of 'justification', which quickly reduces to attempts to distinguish 'rational' and 'irrational' paths to inquiry, which apply not only to the case at hand but to all cases in which knowledge of the truth might be sought. The many efforts, starting with Bacon and Descartes in the early modern era, to define a 'method' that in principle might justify any true belief are the culmination of this line of thought. I regard this fixation on the justification of knowledge claims as involving the exercise of *modal power*, by which I mean control over what people come to think is possible, impossible, necessary and contingent (Fuller 2018: ch. 2). In the *Theaetetus*, the lawyer is made to appear guilty of presenting the contingent as if it were necessary – the 'falseness' of what he says comes not from the falseness of his premises – each of which may be presumed to be true – but their contribution to the validity of his overall argument.

In the *Republic*, Plato had his own signature way of handling the matter, namely, to drive the maximum logical wedge between 'true' and 'false' so that they correspond to what is *necessary* and what is *impossible*. Thus, Plato depicted 'true' and 'false' as contraries. He could then dismiss the poets and playwrights who offered alternative visions of reality to the 'philosophically correct' one as dangerous purveyors of fictions who have no place in his ideal polity. Plato's fingerprints can be found in the recent open appeal to *fear* in the face of uncertainty if, say, American voters failed to elect Hillary Clinton as president or British voters failed to vote to remain in the European Union. 'Either my way or no way', so to speak, was the modal power message. In this framing, Trump and Brexit, respectively, constituted an impossible state of being. Nevertheless, both were selected – and the world has not yet come to an end.

By presenting a state of uncertainty as one of excessively high risk, the American and British political establishments tried to make themselves appear to be the embodiment of rationality and their opponents the face of irrationality. Nevertheless, the voters failed to be moved by this rhetoric, though it convinced the *Oxford English Dictionary* to make 'post-truth' 2016's word of the year. As of this writing, voters in both countries have yet to be persuaded that they made the wrong decision. They may still change their minds – even after the fact, in the spirit of either regret or relief. But it's unlikely that Plato's rhetoric of imminent danger will play a role, given the amount of time that has now passed for US and UK voters to get used to outcomes that had surprised all sides.

In contrast, Bacon, Descartes and other promoters of the scientific method in the modern era have drawn the logical battle lines for modal power more narrowly. For them, 'true' and 'false' correspond to what is *necessary* and what is *contingent* – a matter of contradiction, not contrariety. For Bacon, the art of experiment is ultimately about trying to construct a case in which one of two hypotheses that predict the same thing in every other case ends up failing to do so in the test case. Such an experiment is 'crucial' because it reveals the contingent pretender to necessity. Karl Popper, arguably Bacon's smartest reader since Kant, promoted this strategy to a world view. For his part, Descartes proposed that God's existence ensured that the crucial experiment sought by Bacon and Popper 'always already' exists, so we don't need to worry about an 'evil demon' whose seamless representation of reality might be forever fooling us into mistaking the contingent for the necessary.

Suppose someone happens to have a run of luck predicting people's fate and then becomes the court astrologer. Suppose a certain dance manages to produce rain often and memorably enough and then becomes a tribal ritual. And suppose a social scientist infers a causal relationship from a reliable correlation between two streams of events. In the history of modern philosophy, 'empiricism' and 'induction' have been the terms of art to capture these contingent appearances of knowledge, typically with a stress on people's 'impressionability'. David Hume cast a decisively sceptical spin on these simulacra in the mid-eighteenth century, which was given momentum by increasing suspicion towards 'habit' as a basis of human behaviour by self-styled 'progressive' thinkers. Nowadays cognitive scientists are naked in their antipathy to habit, dubbing it 'confirmation bias'.

Truth be told, 'confirmation bias' is a very post-truth way of talking about not only the sorts of inferences that Bacon and Descartes would regard as merely pseudo-justificatory but also the ones that they would accept as justificatory. The word 'bias' suggests that the inferences concerned are both *motivated* and *restrictive*. These are predicates that the post-truther is happy to work with because they suggest the workings of power over the imagination, which is the only way that any sort of 'truth' is produced. The most natural way to appreciate this point is in the context of *games*.

Speaking in a gaming spirit, all of the contingent 'unjustified' ways of believing what is true constitute 'cheating' because the truth-believers had not played by the rules. But the post-truther is always on the verge of asking: Why play this game rather than some other to determine the truth? To respond by making exclusive reference to the associated benefits delivered by the established game – that is, without reference to its corresponding costs or a cost-benefit balance sheet of alternative games – is to open oneself to the charge of question-begging. To economists, arguably the original post-truthers, this is just a failure to factor in 'opportunity costs'. In any case, the post-truth condition takes very seriously that specific values are embedded in any set of rules, which serve to bias the resulting game towards players with certain skills and dispositions. In short, confirmation bias is inevitable in whatever mode of justification one adopts.

Consider the widespread judgement that homoeopathic treatments are inferior to normal 'allopathic' medical ones for most ailments. When medical scientists conduct the relevant tests, homoeopathy tends to perform quite poorly. Such tests normally focus on eliminating the specific physical source of the ailment. However, homoeopaths complain that this begs the question, since their brand of medicine regards the physician–patient interaction as a constitutive factor in any treatment. Thus they argue that a fair test of their treatments requires the registering of subjects' own sense of well-being, independent of the physical state of their ailments. In response, medical scientists are inclined to dismiss such claims as the 'placebo effect', since for them treatment is primarily about healing a specific part or function of the body rather than the whole person.

What I have just described is a dispute about the rules of the 'medical science' game, in terms of which 'winners', 'losers', 'fair play' and 'cheating' can then be defined. When game-talk was fashionable in philosophy a couple of generations ago, the frame of reference was Ludwig Wittgenstein's (1953) conception of 'language games', which he took to be constitutive of the entire social order. Some (Winch 1958) earnestly thought the idea had profound implications for the very possibility of social science, basically ratifying common sense over 'scientism', while others (Gellner 1959) thought that such 'common sense' was simply a euphemism for Anglophone imperialism. In any case, Wittgenstein thought of 'native' players of a language game as also its referees, who ran into problems only when faced with strange plays or strange players. From this standpoint, homoeopaths don't have much of a chance in the medical science game, since the 'native' physicians hold all the cards. After all, homoeopathy is really about changing the rules of the medical science game, not simply resolving an ambiguous play in the already existing game.

However, Wittgenstein's conception of games wasn't the only game in town. Mathematicians and economists were developing what, after John von Neumann and Oskar Morgenstern (1944), came to be called 'game theory', arguably the most enduring intellectual legacy of the Cold War. It is about constructing games from scratch in situations of uncertainty rather than negotiating the uncertainty in already constrained Wittgensteinian games. David Lewis (1969) popularised this alternative conception for mathematically illiterate philosophers by repurposing the term 'convention' to characterise the task at hand. One philosopher who applied this approach in a striking way to address problems of justice was John Rawls (1971), who argued that a 'veil of ignorance' was needed to resolve potentially irreconcilable differences. Instead of talking about an ambiguous state of play in an already existing game, as game theorists tended to do, Rawls shifted the focus to the rules of the game itself. Thus, he prescribed that everyone must reconstruct their interests in a state of radical uncertainty about their true starting point in the 'constitutional game' that underwrites the society in which they would all live.

The post-truth condition amounts to normalising Rawls's veil of ignorance because there is always everything to play for in terms of establishing the rules of this ideal just society, given that one does not know whether one starts from a relatively advantaged or disadvantaged position. Rawls himself was trying to establish a level playing field with regard to judgements of welfare, with an eye to providing a 'safety net' to the potentially disadvantaged in the envisaged society. However, the same principle could easily be applied to judgements of validity in research. In other words, researchers should decide criteria of validity without knowing whether they begin as 'normal' or 'deviant' with regard to their field of inquiry: allopaths or homoeopaths, evolutionists or creationists and so on. In short, if your knowledge claims might be hard to prove in this new society, what should be the burden of proof that you are expected to bear? Intuitions about the 'neutrality' of the scientific method have historically begun from something like this Rawlsian sensibility – and we lose them at our peril.

Clearly the language gamers and the game theorists are not playing with the same sense of 'game'. Wittgenstein's sense of game drew on the anthropological task of understanding what passes as normal in an alien society *that one doesn't wish to disturb unnecessarily.* That last clause made sense in the late imperial period when there was a strategic advantage to allow the natives conduct their lives in their own ways as long as they were willing to play onside with the imperialists in their games against other imperialists, a situation that became especially clear in the twentieth century's two world wars. That was the *Realpolitik* that gave rise to modern *relativism*. In contrast, game theory served to alert philosophers to the unintended and sometimes perverse consequences of deciding to play one game rather than another in an open-ended situation that combines conflict of interests and obstruction to communication. There is clearly more to play for here than in the Wittgensteinian sense of game – and that is what the 'post-truth condition' is ultimately about.

The shift here reflects the difference made by the Cold War whereby the two main opponents – the United States and the USSR – had each laid claim to the entire world as their potential sphere of influence. Moreover, they made universal claims for their own versions of such pivotal concepts as 'democracy', 'freedom', 'equality' and 'justice'. And these versions drew largely from the same sources in Western political history and philosophy. Yet on their face the conceptions proposed by the two sides were mutually exclusive: one side's version of democracy was the other's version of tyranny. Who to believe? The reason that the Cold War has been often called an 'ideological' struggle is that the battle was ultimately over the frame of reference for understanding these key concepts. Early in the Cold War, the Scottish philosopher W. B. Gallie (1956) spoke of 'essentially contested concepts' to capture the potential of certain politically or religiously charged ideas to stymie rational debate because they are prone to 'incommensurable' interpretations, to recall a fashionable term in the philosophy of that period (Fuller 2000b: ch. 3). However, Gallie still believed that relatively few concepts were essentially contested. Most were more or less governed like Wittgensteinian language games. However, in the post-truth condition, *all* concepts are 'essentially contested'.

Here is a way to measure the conceptual distance travelled from imperialism to the Cold War and beyond. The objective of the imperial game had been to occupy strategically valuable geographical spaces. It was prosecuted very much in the spirit of two-dimensional board games, as reflected in the typical layout of a 'war room'. This horizon emerged from the concerted efforts of leading nineteenth-century mathematicians and physicists to produce accurate maps and synchronised clocks capable of marshalling an unprecedented movement of people and things across a rapidly globalising world. Thus, the space–time matrix became the public face of a world that had turned into one big playing field – and life into one big board game. By the end of the nineteenth century, every self-respecting nation state had to have its 'war room', which by the second half of the twentieth century had to include a computer simulator, which nowadays occupies most of the attention though much less of the space in the room.

Moreover, the effect of gamification of reality extended beyond the geopolitical ambitions of imperialism and world government to how we think about thinking itself. In particular, it promoted the normative standing of so-called linear thinking, namely, the stepwise instructions that are characteristic of the algorithms that execute computer programmes today, including ones that deal in such non-linear 'complex' processes as climate change or business trends. In this way, reality itself has come to be seen as the execution of a programme, a set of possible plays in a rule-governed game, the actual outcomes of which the programmers themselves may not be able to anticipate with any serious specificity because conceptual horizons have become untethered from geographical intuitions.

The background to this development begins with Euclid's presentation of geometric proofs back in third century BC Greece. But as the word 'geometry' implies, Euclid was speaking only about the abstract structure of the Earth as a gaming space. Proper 'geo-gamification' was a largely nineteenth-century effort to turn the Newtonian world-picture into the structure of human understanding, just as Kant had prescribed in the *Critique of Pure Reason*. The crucial step was taken in 1854 to present all thinking in this manner in a book entitled *The Laws of Thought* by the Anglo-Irish mathematician George Boole. It turned conceptual space into gaming space. The field called 'logic' had always been about how we ought to think, but never before had it been presented as Boole did – a version of algebra expressed in proofs. He even applied the method to settle complex metaphysical arguments. Boole effectively invented 'symbolic logic', which set the standard for all reasoning after Alfred North Whitehead and Bertrand Russell's ambitiously titled *Principia Mathematica*, which was published in several volumes in the early twentieth century. Mimicking both the name and style of Newton's own great work, it not only made the reputations of both authors, but for most of the twentieth century it was heralded as one of humanity's great intellectual achievements. Because Whitehead and Russell's *Principia Mathematica* is as boring to read as its Newtonian inspiration, our digital age tends to take its significance for granted.

What is perhaps most taken for granted is that the 'symbols' in symbolic logic consist in what mathematicians call 'functions', which relate 'variables' that issue in a set of 'products'. These functions require data inputs from outside the system in order for the system to do the work of delivering the products, aka outputs. While Whitehead and Russell had demonstrated how all of reasoning could be reduced to a system of simultaneous algebraic equations, there remained everything to play for in terms of the equations that are used and the coding of data as inputs in those equations. It doesn't matter whether the equations or the data are entered by a human or a self-programming mechanical computer. In both cases, symbolic logic made reality more explicitly game-like as a 'universe of discourse', Boole's own phrase for what gamers would call 'field of play'. Note the role of discretion here: the same datum may be cast as positive or negative, typical or atypical and so forth, depending on the system of equations that governs the algorithm in which it figures. In this respect, 'data' function as avatars whose ultimate fates depend on the version of 'Second Life' in which they are inserted. And what we call 'reality' is a fusion of those Second Life versions.

The Cold War raised this 'meta-gaming imaginary' to full self-consciousness. Instead of playing by pre-existing rules of the game, the players basically make up the rules as they go along, such that a set of moves that give one side an initial advantage could end up giving the other side the advantage, if the rules are effectively changed along the way. This is why it is always difficult to know who is ever 'really' winning or losing a war until it is formally over. Welcome to reality as defined by game theory! Lest one think that this is mere speculative fancy, a reasonable historical generalisation is that the bigger the conflict, the more likely that its resolution will be unrelated to its pretext. This may be because the originally disadvantaged party somehow manages to get the upper hand during the conflict. But it need not happen that way. Rather it may simply be that whichever party turns out victorious comes to feel that it must compensate for an excess application of force. This is arguably the sense in which Japan and Germany benefited more from having lost the Second World War than had they won (Fuller 2018: ch. 7).

In any case, the last hundred years of warfare have witnessed the rapid rise in the meta-gaming imaginary. The First World War, an otherwise unnecessary conflict, displayed a combination of aerial combat and relatively advanced field communications that coordinated action on several fronts at once. The result was that the two-dimensional board game vision of warfare went 3-D. Its full force was felt in the Second World War, when aerial combat emerged as the most significant theatre of warfare. At the same time, the fourth dimension of time was being added with radar, which enabled early detection – and hence pre-emption – of incoming threats. However, the Cold War complicated this trajectory, as the advent of nuclear weapons potentially levelled the asymmetries between combatants. Henceforth, even a relatively minor power could acquire the means to destabilise the world order on short notice. This prospect was already satirised in the late 1950s novel and film, *The Mouse That Roared*. And with the 'global war on terror' unleashed by 9/11, cyberwarfare has come to centre stage as the ultimate distributor of destabilisation, now symbolised by the autistic bedroom hacker capable of breaching Pentagon security protocols.

3

THE FATE OF TRUTH, REASON AND REALITY IN THE POST-TRUTH CONDITION

In what I call the 'truth condition', there is agreement on the frame of reference in which key concepts are interpreted. This then allows a specific set of rules to govern a game in which opposing players are disposed to agree on the outcomes and the intervening judgement calls on matters of play. The least coercive way to achieve the truth condition is to get people to believe that our normal modes of social intercourse 'always already' presuppose it. This strategy, which after Kant is often called a 'transcendental argument', may work if people are inclined to believe that their society is generally working so well that most of its infelicities can be resolved through relatively minor local adjustments whereby one or another party changes their beliefs or actions. In that sense, the exceptions serve to prove the rule – the ultimate compliment that a society can pay to itself, as sociology's founder Emile Durkheim, given his understanding of 'deviance', might have said.

However, in the modern period when the exceptions have involved larger breaches of the rules, the scientific method has been mobilised as effectively the 'game of games' governing everything. Thus, scientific experts have been deployed to explain everything from the mental state of a murder suspect to the abortive lift-off of a seemingly functional rocket. But this is easier said than done. To be sure, the dream of an expert-driven truth condition remained strong during the Cold War. The scientific polymath Arthur Kantrowitz (1977) even proposed a literal 'science court', which stayed in vogue for a few years. It would be convened to agree the facts at a given moment that are relevant to a science-based policy decision. He envisaged that such a court would be increasingly necessary as we moved to a science-based world. The US philosopher Hilary Putnam (1979) sublimated this sentiment in his account of science's role as the truth condition's court of last appeal in what he called the 'division of linguistic labour'. Putnam held that in disputed matters, only science can decide what, if anything, we are really talking about.

Perhaps the most sublime expression of this general mentality was US sociologist Daniel Bell's (1960) famous declaration of an 'end of ideology', as scientific thinking and its technological applications increasingly populated our ordinary decision-making environments. Later Bell (1973) spoke of our living in a 'post-industrial society' in which technocracy would become the accepted face of politics. His was a world in which Hillary Clinton would have easily beaten Trump and the UK would have voted to remain in the European Union – because judgements would be made on the basis of accepting what the 'competent' and 'expert' people advise. So close yet so far from what actually happened. What went wrong?

To give Bell his due, 'technocracy' did indeed become 'the name of the game' but it was not practised in the sort of way that he had envisaged. His technocratic utopia had been an updated version of a logical positivist-style vision, one in which ideological differences are ultimately resolved by exchanging inflammatory words for computable data points. But instead, what happened was that the ideological differences between the United States and the USSR were displaced by a game that forced them to play on the same side because both were equally threatened by an overwhelming negative outcome. Of course, I mean the nuclear arms race, which featured the prospect of 'mutually assured destruction'. It became the de facto 'game of games', which ended once the cost of maintaining it bankrupted the USSR. Thus, the original ideological differences between the United States and the USSR were never resolved on their own terms, strictly speaking. This helps to explain the continued unconsummated *Marxisant* yearning among Western academics who are critical of American foreign policy, both then and now.

What Bell ultimately underestimated – at least from a post-truth stand-point – is that the Cold War shifted the modal balance of power from con-tingency back to impossibility, from Bacon back to Plato. Thus, the fear of mutually assured destruction focused minds on both sides of the Iron Curtain on 'thinking the unthinkable', to recall Herman Kahn's (1962) reso-nant phrase from the period. It enabled an unprecedented global alignment of scientific and political authority – that is, until one of the key combatants found that it was no longer economically sustainable. In our own day, many would see the prospect of a climate apocalypse – or perhaps more to the point, a pandemic – as the new truth condition whereby all sides come together to avoid the same negative outcome. Perhaps all we need is the right game to make this happen. But with the end of the Cold War, many of those technocrats in whose hands Bell had invested the fate of humanity have gone rogue and become 'hackers' (Wark 2004).

The post-truth condition reduces reason and reality to matters of gamesmanship. They boil down to the standard of personal judgement. What is the standard by which you decide that something is right or wrong, true or false? Does the standard lie inside or outside of you? If you think it lies inside of you, then you're bound to be a *nonconformist* – someone who wants to win by changing the rules of the game because you believe that the game as currently played is unfair. But if you think it lies outside of you, then you're probably a *conformist* – someone who wants to play – and win – by the rules of the game. The conformist is wedded to the truth condition, the nonconformist to the post-truth condition.

Plato's singular genius as a philosopher – working from a database no greater than the Greeks' understanding of their own history – was to recognise that nonconformity is the normal state of affairs and that conformity is something that requires construction and maintenance. Indeed, he was the original social constructivist. His strategy was to promote the idea that one game trumps all the other games – and that everyone has it within themselves to play that game. That game – the basis of the truth condition – may go by the name of 'reason' or 'reality'. In the spirit of the non-coercive 'transcendental' strategy mentioned above, Plato wanted people to believe that they already know this ultimate game but they need to be reminded of its rules and then they will see how to play it properly. He thus sought to achieve conformity through the disciplined application of memory, in terms of which the external world appears as offering cues or prompts to think and do the right things.

Plato's point can be glossed as being about either knowledge or power. When it's about knowledge, it involves getting clear about what one knows and doesn't know. When it's about power, it involves getting clear about what one can and can't do. Plato articulates this strategy in the *Republic*, in which the ideal society is based on people's sphere of activity being proportioned to their capacity for knowledge. Those who know more can do more: knowledge is power. The person who knows the most and can do the most is the 'philosopher-king', in which the Greek word for 'king' (*Basileus*) should be understood in the spirit of the English phrase 'pillar of the establishment' whereby the significance of the position is recognised only if the structure that it upholds is recognised as well. In short, the philosopher-king has the capacity to stabilise the game that he would have everyone else play. In the modern period, this capacity for stability becomes depersonalised as the 'state'.

But Plato's point about proportioning power to knowledge also applies reflexively. In other words, those who know more about the limits of their own knowledge are capable of maximising their power within those limits – or at least make it less likely that they will suffer needlessly if not implode altogether. Hegel famously captured this sensibility as 'freedom is the recognition of necessity', but it had been made into a cardinal virtue by the Stoics, a school of philosophers who flourished after the fall of Athens and throughout the Roman period. Their concept of *duty* – the 'ought' – was carved out of this reflexive understanding of 'knowledge is power'. It amounts to a form of self-exploitation whereby one neither exceeds the limits of one's capacities nor – and perhaps more importantly – fails to meet them. When we speak of people 'underachieving' or 'not living up to their potential', it is about failure to meet this core sense of duty. In short, all duties are ultimately duties to oneself.

In the modern period the concept of duty underwent two interrelated changes: first, people came to be seen as increasingly similar in capacities, and second, one's willingness to exercise those capacities even in the absence of any obvious reward became the cornerstone of what we now call 'autonomy'. And since Kant, this sense of autonomy has been a strong moral and legal indicator of personhood in humans – and possibly non-humans. It was also inserted into a renovated framework of the knowledge–power nexus, one in which 'science as a vocation', to recall Max Weber's (1958) resonant phrase, came to acquire much of the authority that Plato had reserved for the philosopher-king in the *Republic*. In both cases, the 'freedom' that scientists exercise in their research and teaching is made possible by their simultaneously upholding the background conditions in which they operate. That makes what they do a truly 'rational' form of freedom. Thomas Kuhn (1970) later drove home the point in a particularly brutal yet popular way. Although all scientists know that their work is likely to be superseded, that does not lead them to adopt an anarchic approach to inquiry because they also know that whatever freedom they have to make a difference now depends on their having honoured the legacy of their predecessors. They have effectively tied themselves to the mast.

One can hear the echoes of the *Basileus* in all this. But the post-truth thinker hears something else as well. It is the military idea of *Spielraum*, the 'room for manoeuvre' or 'wriggle room' whereby 'rational freedom' means the opportunity to maximise one's own position in a clearly defined state of play. Although people often regard the comparison of war to games trivialising, modern warfare is probably best understood as an amplified version of gaming along the dimensions of resources, participants and stakes in the outcome. Common to the dynamics of warfare and gaming is the idea of playing within the rules yet radically changing the state of play. A small opportunity can turn into a big advantage, while a big opportunity can end up in disadvantage. Weber and later Karl Popper (1957) spoke of the 'logic of the situation' facing any social agent very much in this spirit – as part of a general argument against any overarching sense of 'determinism' at the historical or cultural level.

4

CAPITALISM, SCIENTISM AND THE CONSTRUCTION OF VALUE IN THE POST-TRUTH CONDITION

Capitalists approach the world in a 'post-truth' frame of mind because in a market, standards of value are co-created with the things valued by those standards. Values do not pre-exist the price system. On the contrary, they are its most enduring product. Markets are games whose rules are perpetually under construction by the players from whom 'winners' and 'losers' are selected. This is what 'classical liberals' such as Friedrich Hayek mean when they extol the 'self-organising' nature of markets. The rules governing a market at any given time are no more than conventions, whose continued enforcement depends on the fate of those subject to the rules. Indeed, the increasing portion of corporate budgets dedicated to marketing amounts to an ongoing campaign to stabilise or change standards of judgement to corporate advantage. In this respect, capitalism is the ultimate 'social constructivist' project which helps to explain why the US sociologist most closely associated with social constructivism, Peter Berger, was also the field's staunchest defender of capitalism as a value system in the late Cold War period (Berger and Luckmann 1966; Berger 1986).

It is worth recalling that when Count Saint-Simon coined 'socialism' in the early nineteenth century, he meant it as complementary to capitalism in the sense that Plato's truth-regime was complementary to the post-truth regime that had existed in an Athens dominated by sophists. In both the ancient and modern cases, at stake is the productive channelling of individual freedom: the sophists and capitalists would run free with minimum but clear constraints, presuming that the resulting benefits would outweigh the harms along the way – and because of the clarity of the constraints one could pre-empt or defend against similar harms in the future. For their part, Plato and the socialists reckoned in reverse, resulting in their restricted – they would say 'focused' – sense of freedom that would aim to eliminate the prospect of harm altogether. But where that is not possible, the aim would be to reinterpret the harm as a larger benefit in disguise. Two early Cold War books with resonant titles, Karl Popper's (1946) *The Open Society and Its Enemies* and Friedrich Hayek's (1952) *The Counter-Revolution of Science*, got the measure of this difference in sensibility – and basically sided with the more 'liberal' sophists over the more 'socialist' Plato. Isaiah Berlin (1958) famously cast this dichotomy in terms of 'negative' versus 'positive' liberty.

When capitalists are accused of reducing *value* to *price*, they gladly plead guilty because for them price simply reflects the relative ease with which people are willing to surrender something they have for something they don't have. Value is ultimately about the cost of replacement in an economy whose fundamentals are in principle always under construction. The way to assess the true value of something under capitalism is to see the price at which a transaction goes through. This helps to explain the role that *auctions* have played in spreading the capitalist mentality across all of society, ranging from ordinary commodities to artworks to even human lives. The last context immediately brings to mind slavery. However, as Marx famously emphasised, a perverse auction also operates for wage labour in modern industrial society, as long as workers remain unorganised in what is effectively a buyer's market. In that case, workers are forced to bid their wages *down* to make themselves more attractive than their fellows to potential employers. David Ricardo, the classical political economist on whom Marx had most clearly in his sights, regarded this tendency as an 'iron law'. It was in this spirit that German sociologist Werner Sombart (2001) coined 'capitalism' in 1902 to name what had become a general world view.

For Sombart, capitalism marked the breakdown of traditional, typically hereditary social relations based on 'status', a word whose Latin root suggests standards that are fixed independently of the people bearing them at any given time. Thus, from early nineteenth-century Romanticism onwards, capitalism's characteristic cultural expression has been strongly 'anti-classical', where 'classic' is understood as setting a 'timeless' standard, as the Greeks and Romans had supposedly done. A key moment in this transition occurred in 1850, when the influential French critic Charles Sainte-Beuve repurposed the word 'classic' (*classique*) to refer to a *contemporary* work that anchors taste and judgement in a field, in a way that we would now regard as 'fashionable' or 'trendy'. Unlike the ancient classics, these modern classics are designed less to be venerated from a distance than to be experienced and then superseded by later artists who take them as challenges to over-come, perhaps through acts of incorporation that run the risk of 'plagiarism'. Indeed, the late Yale literary critic Harold Bloom (1973) famously wrote of the 'anxiety of influence' from which poets suffer as they try to reverse engineer the works of their predecessors in order to redeploy them to still greater effect. Indeed, their own 'genius' depends on readers forgetting those giants on whose shoulders they stand as they push the giants further into the ground.

All of this is very much in the 'creative destructive' spirit of what Joseph Schumpeter (1942) later called 'entrepreneurship' to characterise the corresponding innovative sensibility that marked the Industrial Revolution. It follows that there are no standards without standard-bearers, and in principle anything or anyone can be a standard-bearer, if the achievement effectively generates new markets that reflect the achievement's dimensions of success. You know you have set a new standard when others are trying to outrun you in a race that you are recognised as having started. Such is the calling card of successful entrepreneurship. In that respect, one doesn't even know what a market is truly 'about' until one sees what traders treat as alternatives, which in turn relates to some perceived sense of functional equivalence in satisfying a set of needs within a budget. And so, 'the name of the game' comes into view. However, once the market is disrupted by a newcomer, all the calculations may shift as people realise that they're playing a different game – and that the dominant values have changed.

Under capitalism, a market is not a place in a town where people trade specific goods and services. Rather 'the market' is a concept – a second-order entity – with the potential to comprehend all social relations, as the roles of 'buyer' and 'seller' are adopted under various negotiating conditions for purposes of enhancing one's own overall advantage. Thus, the natural con-dition of market formation – namely, that people routinely have too much of some things and too little of others – can be manufactured at will. It becomes what sociologists call a 'script', which generates a particular game, including positions and rules of exchange. This is the frame of mind in which Sombart turned the circulation of capital into an '-ism', 'capitalism', thereby rendering it an ideology that might be explicitly adopted and defended – indeed, something around which a political strategy might be organised. Sombart's great contemporary and rival, Max Weber, got the measure of this as part of the dynamics of modern parliamentary politics. Here 'party platforms' function like today's computer-based 'game platforms' in that both prefigure the space in which policies can be proposed and decisions taken. Politics then becomes competition for control over the game being played – and elections constitute the market in which the relative merits of these games are tested and decided.

In contrast, Marx himself tended to treat capital as an impersonal, albeit universal, factor of modern production that in the long term resulted in both workers and their employers acting against their own self-interest, eventuating in total collapse of the system. Thus, when Marxists began informally referring to 'capitalists', the term was used derisively for people who were suffering from a special self-aggrandising form of false consciousness. Even today Marxists continue to use the label 'capitalist' pejoratively, while oblivious to the fact that most of the intended targets gladly embrace the label. It would seem then that Sombart has had the last laugh.

This brief look at how 'capital' acquired its '-ism' sheds light on science's historically fluid relationship to its evil twin, *scientism*. Many of the people accused of being 'scientistic' – and they include both professional scientists and policymakers – have seen themselves as simply engaged in the extension of scientific principles to new domains as part of science's rationalising mission. When Hayek (1952) popularised 'scientism' as a cover for totalitarianism, he was concerned with protecting the brand name of 'science' in the face of self-styled 'social engineers', who invoked science to inhibit people from exercising their freedom. While Hayek was especially suspicious of socialists, the more general tendency to defer to experts in modern complex democracies fell under his original concern (Levy and Peart 2017).

'Scientism' has also come to mean the expansion of science into domains where it only serves to prop up the prejudices of particular scientists, as when Richard Dawkins claims that evolution implies atheism. But increasingly 'scientism' is used for practices and policies endorsed by scientists that have had adverse consequences for vulnerable groups in today's society. This sense of 'scientism' was clearly on display in the sesquicentennial issue of one of the world's leading scientific journals, *Nature* (Comfort 2019). To be sure, those accused in this context of 'scientism', such as eugenicists, were not necessarily malicious, naïve or ignorant vis-à-vis these consequences. Nevertheless 'scientism' is used to cast an absolute moral judgement that the practices and policies themselves – and perhaps even the ideas that informed them – are bad, if not 'evil' in some metaphysical sense. Moreover, it is suggested that historical knowledge might help ensure that such 'scientism' does not happen again.

Unfortunately, this would be a misuse of history to oversee the future. What counts as 'good' or 'bad' in scientific practice or science-based policies can be understood only in retrospect because they are ultimately effects of the same sort of causes. It is the difference in our attitude towards the outcomes that makes one set 'good' and the other 'bad'. The field of science and technology studies converts this point into methodology, known as the 'symmetry principle' (Bloor 1976; Fuller 2018: ch. 3). There is no principled way to distinguish the 'scientific' (good) and the 'scientistic' (bad) until we examine our own relationship to the phenomena in question. After all, the same nuclear physics that would teach the world the secret to cheap and clean energy had also taught the world the most efficient means of destroying itself. The same chemistry that gave us artificial fertilisers also gave us poison gas, the same biology that first backed eugenics now backs gene editing – and the list goes on.

Jean-Paul Sartre captured the ethical side of this situation as the problem of 'dirty hands', which stressed how the 'good' and the 'bad' are intertwined in any causal account of how things came to be as they are. Whatever moral clarity we seem to have presupposes our having identified with only some of the entwined factors but not others. Thus, the undeniably bad stuff that has come from science is accepted as 'collateral damage' from the good stuff that outweighs it – what after Aquinas is called the 'doctrine of double effect' (Foot 1978: ch. 2). In any case, it is a decision that we take and for which we are then ultimately held responsible. Indeed, Existentialists go further: it is only when we deem that a particular action is 'good' or 'bad' (or a particular belief is 'true' or 'false') that we discover who we really are. In general, we should expect that these two interlocked judgements – who 'we' are and what counts as 'good' and 'bad' – will change over time. It recalls the value construction that we originally identified as endemic to capitalism. Like a new product brought to market, the moral and epistemic character of any event is indeterminate at the time it happens. It contains a range of possible pasts and futures that are only settled after the fact, but which in principle at least may be reopened later. In this respect, science is a quantum phenomenon – and 'scientism' is its observer effect.

Consider 2009's 'Climategate', the product of a Freedom of Information request by a UK journalist for e-mail exchanges circulated by a top international team of climate scientists. They revealed, among other things, efforts to textually and visually represent climate data in the strongest possible light so as to preclude any scepticism about the magnitude of the climate change evidenced. What might have been judged a methodological malfeasance on the part of the climate scientists – spinning the data – was in the end treated as operating within the discretionary bounds of methodologically sound behaviour. This undoubtedly had to do with who the agents were and what they were saying – both understood in terms of the current state of play in the climate change debate. For now, the verdict on Climategate is that it was proper 'science' and not improper 'scientism' done in the name of promoting a particular political agenda. However, it is easy to imagine a reversal of this judgement in the future, if it turns out that the current global climate crisis has been more generally misjudged.

To put the point in sharpest relief, eugenics came to be seen as morally untouchable only after the 1946 Nuremberg trials over the Nazi atrocities done partly in the name of eugenics. In its wake, the 'civilized' world was systematically inhibited from learning from the science done under Nazi aegis, not least by a prohibition on the citation of Nazi scientific research. Complementary to these efforts to quarantine inquiry has been an unprecedented increase in moral safeguards on research with human subjects. The ironic long-term result has been a lack of both epistemic and ethical preparation for the 'transhumanist' era on which we are embarking whereby 'gene therapy' and 'gene editing' amount to a new and improved version of eugenics, this time presented without the historically offensive Nazi backstory (Fuller and Lipinska 2014: chs 3–4).

5

PUBLIC RELATIONS AS POST-TRUTH POLITICS, OR THE MARKETISATION OF EVERYTHING

Public relations is what a true Platonist means by 'politics', except that Plato would have the state do it rather than having it turned into a for-profit business operating in a competitive environment. Indeed, were Plato teleported to the twentieth century, he would probably fail to see the irony in Aldous Huxley's *Brave New World*, which depicts a society largely governed by state-run advertising campaigns. Where Huxley intends dystopia, Plato would understand utopia. What Plato would like about public relations is its underlying principle, one that characterises marketing in general: it presumes that A knows B's interests better than B, but instead of simply telling B what those interests are, A designs environments that enable B to discover them in his or her own way. This is the strategy that Plato portrays Socrates as adopting in the *Meno* to demonstrate that the slave boy knows much more geometry than anyone had ever imagined. It would be easy to understand Socrates as having simply manipulated the slave boy to give the right responses and then flattered him by claiming that he had previously lacked the opportunity to display his capacity for geometry. But what if Socrates is seen as having indeed succeeded in tapping the slave boy's latent capacity for geometry, which had previously remained untapped, perhaps because his life had been preoccupied with matters that inhibited the full realisation of his potential?

In his classic 1928 work, *Propaganda*, Edward Bernays, Sigmund Freud's nephew and the father of public relations, pressed the second interpretation to maximum effect. Bernays believed that public relations could raise the quality of democratic decision-making in a mass society. The historic problem with democracy had been its default tendency for people to follow the crowd – or someone who claims to speak for it, a 'dictator' in the modern sense. Bernays wrote when for the first time the entire adult population could vote in several countries, and many feared the worst. Nevertheless, Bernays argued that an increasingly liberal society should encourage people to be exposed to more opportunities for decision-making, including to explore undiscovered aspects of their being in the name of personal empowerment. It should thus come as no surprise that the great US psychologist of 'self-actualisation', Abraham Maslow, spent his final years in the 1960s as a high-end marketing consultant (Maslow 1971, 1998).

Public relations was originally pitched to those worried that modernity had simply replaced the conformity imposed by birth and custom with the conformity imposed by the recently empowered masses. Bernays promised to reconcile democracy and liberalism for the greater good of society by proactively enabling people to express their full capacities. This is the spirit in which marketing campaigns invite the consumers to explore the hidden recesses of their imaginations. In practice, Bernays's proposal meant the endless manufacture of decision spaces for people to choose between policies, products, whatever. After all, it makes no sense in capitalist society to manufacture products endlessly unless there are also endless opportunities to buy them. Thus, Bernays's project is fairly seen as the 'marketisation' of everything, including people's reputations. And of course, he succeeded, turning public relations into the multibillion dollar industry that it is today. Indeed, for many years now, capitalist efficiency gains have driven down production costs to such an extent that corporate budgets are more devoted to marketing than making products. The same trend can be seen in politics, in which the size of a candidate's 'war chest' has become the best predictor of electoral success – certainly a better predictor than the candidate's actual political track record.

Whatever else one might say about this development, it puts an ironic spin on John Kennedy's comment on the 1962 Cuban missile crisis, 'The cost of freedom is always high', since *cost* implies the sum total of all you pay, even if not upfront, as in such 'negative externalities' as environmental degradation which are often only discovered after the fact. However, public relations firms are capitalist enterprises that charge upfront. Thus, Kennedy would have been more correct to say, 'The *price* of freedom is always high'. And why does freedom carry such a heavy price tag? Because it takes a lot of work both to mine and craft the information needed to create proper decision spaces for people to think that there is something at stake among alternatives. Nowadays much of this is discussed in the context of 'data analytics' firms that collate people's spontaneous 'digital footprints' to identify pressure points in targeted campaigns to shift a critical number from their default positions (Thompson 2020: ch. 3). To be sure, these firms are associated with 'Russian' or 'Chinese' interference in Western elections. Nevertheless, *Wired* magazine founder and all-purpose Silicon Valley guru Kevin Kelly (2014) has adopted a somewhat more sanguine attitude, focusing on efforts to make people effective shareholders of the firms that are routinely mining and crafting their data, an arrangement that he has gamely dubbed 'coveillance'.

At the most general level, marketing is a form of emotional alchemy that plays on the axis that defines the emotional texture of the post-truth condition: *fear* versus *hope*. What these two emotions share is an attitude to a non-existent place in time called *the future*. They treat the present instrumentally as a moment of play in some longer game. In that respect, whatever counts for truth now is at least to some extent discounted. This explains post-truth's flexible attitude to *facts*, which are basically treated as randomly injected events that in principle can be turned to any player's advantage. In that case, fear inspires consolidation and self-protection against some anticipated threat to what we imagine ourselves to have achieved ('hard won facts'), whereas hope inspires openness and boldness towards a future in which we believe there is everything to play for ('facts yet to be revealed'). The former is oriented towards what we might lose, the latter towards what we might gain, by shifting our current position: the precautionary risk-avoiders vs. the proactionary risk-seekers, the *lions* versus the *foxes*, in Machiavellian terms (Fuller 2018).

Marketing plays on both attitudes towards uncertainty by instilling a fear that is then leveraged into hope, all designed to propel a desired course of action. First, consumers must come to think that they are missing something crucial to their lives before they construct for themselves a desire that might be then satisfied by a particular product. Thus, marketing is about inducing a sense of privation in the consumer, which then motivates one to make a purchase, from which one receives a renewed sense of completion in one's life: from fear to hope to salvation. Moreover, this process may be repeatedly triggered, say, through planned obsolescence whereby the manufacturer asserts that the original purchase no longer satisfies the consumer's now more discerning needs, respect for which requires the manufacture of a new product that merits the consumer's consideration. For her own part, the consumer may have yet to register the insufficiency of the old product but a successful marketing campaign aims to induce just the right amount of misgiving to motivate a new purchase. Nowadays members of the 'Apple community' of smartphone users are routinely played in this fashion. From this standpoint, politics is much easier to market, since regular elections in a democracy amounts to planning the obsolescence of policies and politicians into the political system. Thus, the fear generated by the question, 'Where do we go from here?' is routinely induced.

The underlying psychology to all this can be explained by the Gestalt principle of *Prägnanz*. Think 'pregnant with meaning'. Here one should imagine a classic Gestalt experiment, in which subjects are presented with an inchoate perceptual stimulus, plus background – visual or verbal – that enables them to discern a specific object. Thus, in the famed 'duckrabbit' experiment, the same ambiguous figure may be understood as depicting either animal, depending on the cues given to subjects. The subject's efforts then provide the requisite sense of completion associated with *Prägnanz*, a state of cognitive or material satisfaction. The apparent arbitrariness with which the subject encounters the object provides the occasion for a sphere of meaning to be constructed that encompasses the two, reconfiguring them as parts of a common whole, hence 'Gestalt'. Generally speaking, in the face of uncertainty, people are inclined to fill in the missing information with their own concerns which they then use to justify the decision they have taken. People don't simply want to see things as they appear to them; hence their dissatisfaction with an ambiguous stimulus. Rather, they insist on resolving that ambiguity as an object to which they have some specific relationship.

This helps to explain the 'sweet spot' in marketing when the consumer sees a particular vehicle on the showroom floor and says, 'This car was made for me!' And so the deal is sealed, and the two parts – the customer and the car – are rendered whole. Now consider what preceded this moment, especially the highly romanticised images of freedom and control that have been associated with automobile advertising for most of the industry's history. Such ads manage to tap into matters sufficiently close to consumers' deep self-understanding that they are capable of discounting the hype of the ads to purchase the vehicle. Most men realised that they wouldn't get the sexy woman reclining on the car or make those hairpin turns on the Swiss Alps.

But they would feel better about themselves and probably get around more easily.

Bernays was the grandmaster of this art. It plays on creating what social psychologists call 'cognitive dissonance' in the audience. In his first major public relations campaign, Bernays persuaded Americans to sign up to the First World War, which was shaping up to be the bloodiest conflict in human history. Nevertheless, it was being fought entirely on European soil. The United States had not been formally attacked, though a stake had been created with the sinking of the *RMS Lusitania* in 1915, in which a German submarine torpedoed a British ocean liner, a tenth of whose casualties (128) were Americans. While the event was widely portrayed as 'collateral damage' from a war that was being conducted at an unprecedented scale, President Woodrow Wilson was keen to use it as a pretext for US military involvement, not least to promote his own 'Progressive' view of America's global ascendancy. Interestingly, Bernays's campaign of persuasion did not focus on revenge, which could have escalated the already high level of jingoism associated with the war in Europe. Instead he took the high road, portraying American entry as a show of gratitude to the Europeans who throughout the nation's history have risked their lives to make the United States the bastion of democracy and prosperity that it had become by the early twentieth century. Thus, two events that were previously seen as independent of each other – the security of the United States and the security of Europe – were now recast as interdependent, which in turned generated a psychic tension. The resolution of that tension helped to normalise US overseas interventions for the rest of the twentieth century, resulting in America's becoming the 'world's policeman', a self-image that has really only been in decline since Obama and Trump.

There are two complementary historic precedents for public relations operating in the Bernays style. One precedent is the philosopher-mathematician Blaise Pascal's famous seventeenth-century argument (or 'wager') for the existence of God which hypothetically threatened the reasoner with the prospect of not believing in God in this life and then discovering in an infernal afterlife that the deity does indeed exist. Thus, belief in God is justified because, while not strictly provable, it nevertheless operates as an insurance policy against the anxiety produced by the thoughts of eternal damnation that Pascal had planted. Faith turns out to be a price worth paying for overcoming an important – albeit contrived – obstacle to leading a fulfilling life in the here and now. And even if God turns out not to exist, at least you will have been a good person in the only life you will have led. The second precedent is the idea of inoculation which, in the late nineteenth century, had metaphorically migrated from the immune system of the individual body to the collective mind of the body politic. The shared idea is the production of an artificial prompt (i.e. a vaccine) that activates a latent capacity (i.e. antibodies) that continues to respond, perhaps even more strongly, when natural versions of the prompt (i.e. the disease) occur in the future. Intellectuals have often assumed this role in modern society. They render visible simmering discontents and structural contradictions by portraying them as 'challenges' for society. Perhaps too often, the intellectuals have themselves ended up becoming the messenger shot for delivering the message (Fuller 2009: ch. 3). But perhaps in light of COVID-19, they will be seen in their proper light as the vanguard of 'herd immunity'!

A hundred years ago, the Anglo-Irish playwright George Bernard Shaw got the full measure of this second precedent in *Back to Methuselah*, a series of five plays where he proposed a 'homoeopathic' approach to education. By this he meant that one learns the truth by being routinely exposed to 'white lies', strategically pitched half-truths that serve the speaker's interests but challenge the hearer to discern the true from the false, in terms of both what is said and unsaid, so as to serve the hearer's own interests. This is rather different from Socratic pedagogy, which is normally portrayed as a coaxing process to render the implicit explicit, as in the slave boy's display of geometric knowledge in the *Meno*. In contrast, Shaw wanted education – like the theatre more generally – to provoke people into knowledge, which involves their taking action as they are awakened from ignorance and error. Nowadays we would call it 'consciousness raising'. It assumes that we are born into a problematic world, which extends to the evidence to which we are routinely exposed – including those from whom we might learn. Bernays believed that capitalism routinely placed consumers in this position, as an increasing number of voices and products vied for their votes and custom. It was here that public relations provided guidance. Taken together the two historic precedents on which Bernays modelled public relations issue in the following existential horizon: our generic and largely unconscious desires are elicited by the likes of Bernays in ways that serve to focus our expectations, effectively 'pre-adapting' us to things that aim to satisfy those trained desires. In this way, demand is manufactured or, as Bernays himself put it, consent is engineered.

6

THE NEW YORK TIMES GETS THE POST-TRUTH TREATMENT

The American Philosophical Association gives annual public philosophy writing awards. In 2019 one of them went to Bryan Van Norden (2018) for a piece originally published in the *New York Times*'s philosophy column, *The Stone*. The spirit of Van Norden's thesis is captured well in its title: 'The Ignorant Do Not Have a Right to an Audience'. His target is the media's allegedly supine attitude to our post-truth condition. And what he attacks under the rubric of 'post-truth' is what I defend – namely, the tendency to contest not only the truth or falsehood of knowledge claims but also the terms on which the matter is to be decided. This opens the door to multiple, potentially incompatible standards of evidence, which together spawn what Donald Trump counsellor Kellyanne Conway has memorably dubbed, 'alternative facts'. Van Norden's critique takes no prisoners: he excoriates even such stalwarts of modern philosophy as Rene Descartes and John Stuart Mill for promoting a naively universalist approach to rationality that in practice enables charlatans and cranks to corrupt the public sphere. And it is certainly easy to understand Van Norden's frustration since the post-truth condition puts his way of thinking on the defensive. Nevertheless, his argument stands in the long tradition of philosophers trying to redress an imbalance by imposing their thumbs on the scales – this time, to promote what Van Norden calls 'just access', a principle that would deny the 'ignorant' any entitlement to a public hearing.

Van Norden's intervention recalls the desperation of that influential school of modernist philosophers, the logical positivists, in post-First World War Germany. It was a time of political instability brought on not only by the draconian settlement of the Treaty of Versailles but also by the expressive pluralism unleashed by the Weimar Republic's own democratic constitution. The full ideological spectrum from extreme Right to extreme Left were jockeying for position: Who to believe? The positivists responded by conjuring up various 'criteria' designed to distinguish not only 'true' and 'false' beliefs but also 'meaningful' and 'meaningless' statements. On closer inspection, these criteria were purpose-made to exclude all the religious, political and metaphysical doctrines that they regarded as contributing to the ambient instability. For good measure, the joker in the positivist pack, Karl Popper, provocatively added some fashionable 'progressive' movements – psychoanalysis and Marxism – to the mix of what he called 'pseudoscience'. Perhaps the philosophical high point in this episode was Rudolf Carnap's (1959) takedown of Martin Heidegger's understanding of 'negation' in an early issue of the positivist journal, *Erkenntnis*.

However, it didn't take long for philosophers to start questioning both the feasibility and desirability of the criteria on offer. Nevertheless, these criteria were applauded for being 'public' in the crucial sense that they were readily comprehensible, as they made reference only to observable events and checkable forms of reasoning. It was therefore easy to judge the fairness of how the criteria were applied. This in turn made the criteria vulnerable to charges that they unfairly targeted certain practices. It led Popper and especially his followers to relax their conception of 'criteria' so as to disregard both the provenance and the content of a knowledge claim as relevant to its validity, just as long as the claim is regularly subjected to stringent tests.

In striking contrast, Van Norden's argument is of a piece with our post-truth times. His fierce rhetorical posture, like that of a maimed lion, dispenses with criteria altogether. Instead he resorts to innuendo in the form of a rapidly delivered inventory of intuitions that is designed to conjure an image of how the world has deviated from the judgement of 'right-minded' people. The result is a philosophical version of dog whistle politics, since it is unlikely that an explicit formulation of Van Norden's intuitions as criteria – the 'hidden algorithm' that generated Van Norden's article, if you will – would go beyond codifying the prejudices of the average reader of the *New York Times*.

Beyond the obvious point that any professionally endorsed 'public philosophy' should do more than simply confirm the prejudices of the average *New York Times* reader, its normative strictures should also reflect the vast changes in the media ecology that have occurred over the past half-century.

Van Norden's failure to provide an independent definition of 'the public' suggests that he may be begging the question on what constitutes a 'just access' policy. In particular, his central claim that freedom of expression does not entail a right to an audience presupposes that access to an audience is at least in principle controllable because it is finite. To be sure, the cognitive capacity of any given audience member remains finite, but starting with the cable television revolution in the 1970s, it has become increasingly difficult for an authority – be it the state or the mass media – to control who attempts to gain access to an audience.

Van Norden implicitly trades on an obsolete Enlightenment-based media model in terms of which it made sense to speak of 'the public'. It relied on both the state exercising a monopoly over broadcasting licences and the market ensuring that the entry-level costs of becoming a broadcaster are relatively high. The idea was that a well-educated, well-resourced and well-behaved vanguard would be positioned to channel an emerging democracy towards effective collective decision-making. The model was conceived in the world of the printing press but it was extended to the mass media. The BBC has perhaps remained truest to the model's ideal of 'nurturing' an informed public. However, the annual BBC licence fee – a universal charge of the equivalent of $US200 – has stretched credulity in an increasingly commercialised and pluralist media environment.

Indeed, the past fifty years have drastically altered the dynamics of the situation. Ever since the invention of the printing press, state authorities have found it tempting to adopt a permissive licensing policy in the hope of increasing tax revenues through media-related profits, notwithstanding its potential for generating social unrest. However, in the past, the people who wanted to communicate and the people who owned the means of communication tended to be different. This in turn invited the broadly 'editorial' activities that interest Van Norden, ranging from outright censorship to a more bespoke practice such as the peer-based 'gatekeeping' that characterises academic publication. The difference today is that social media has shrunk the distance between platform and content providers. In principle, anyone can start their own newsfeed or video channel and simply let the market determine its fate. Surprisingly, Facebook and other social media giants have been blindsided by this development, even though they are largely responsible for it. Moreover, their confused response in the face of public outrage and political scrutiny hasn't helped matters. On the one hand, they claim to be mere platform providers and not content providers; on the other hand, they still want to appear to be responsible stewards of the media ecology. Not surprisingly, they have failed to persuade lawmakers in the various countries where they operate.

In contrast, whatever one makes of the politics of the late Andrew Breitbart, he got the measure of the situation right. He responded by strategically re-appropriating a concept close to the heart of the 'cultural Marxism' that he otherwise demonised. I refer here to Antonio Gramsci's notion of the 'organic intellectual' – that is, someone capable of generating an entire world view from a specific position in society (Breitbart 2011: ch. 6). This is how to make sense of former Breitbart News chief executive and Trump strategist, Steve Bannon, and the 'alt-right' movement that was spawned once this modus operandi was converted into a newsfeed algorithm. How exactly one identifies the 'specific position in society' associated with the 'alt-right' is an interesting question in its own right, but the point here is that such things *can* be manufactured.

Do recent changes in the media ecology mean that no justice is to be found? Not necessarily. However, Van Norden's 'just access' principle would need to be re-invented in a world where the name of the game is competitive advantage. His 'right-minded' judgements about the news do not sell themselves to an audience that is not already receptive. The good news here is that receptiveness can be cultivated. One way is by promoting the idea that these judgements have been 'curated', in the expansive sense that has become central to digital culture. Thus, the emphasis would be placed not on the judgements themselves but on the second-order fact that a wide range of alternative judgements have been considered and rejected along the way. Such a news source would be trusted less for its ideological consistency than for its high level of discrimination. This strategy reflects the deeper philosophical point that markets tend to turn problems of epistemology into ones of aesthetics.

The historic motto of the *New York Times* – 'All the news that's fit to print' – expresses just this sensibility. The slogan dates from 1896, when the paper's new owner, Adolph Ochs, wanted to distinguish the struggling *Times* from titles owned by the US media moguls William Randolph Hearst and Joseph Pulitzer. For them journalism provided a campaigning platform along ideologically predictable lines – of the sort that is still practised by mass circulation tabloid papers. In contrast, the *Times* would not be so predictable but it would be 'measured' in a sense that over time became the *Times* enshrined its style of journalism through a self-archiving policy that provided a periodic index of all the topics it covered, including follow-up pieces. By 1913 US librarians were dubbing the *New York Times* the 'newspaper of record'. It continues to make the paper of considerable value to researchers who regard journalism, in the words of Phil Graham, the former publisher of the *Times*' arch-rival the *Washington Post*, as the 'first rough draft of history'.

However, in a market environment, the initial connection to the consumer is at a more sensory level, on the page or on the screen. Apple smartphones may increase their apps indefinitely but the phone itself still needs to look appealing to potential users. In this context, the most influential US journalist of the twentieth century, Walter Lippmann, recommended a rhetoric of 'objectivity' in which the journalist appears knowledgeable of the world's complexity while presenting it in emotionally neutral language. This ethos persists in the self-presentation of mainstream radio and television news broadcasters, as the *New York Times* has traditionally stood for it in the print medium. While Lippmann may seem to have been motivated by a desire to avoid the appearance of bias, he was in fact a student of Plato. Indeed, he was part of the same public relations campaign as Bernays to get the United States into the First World War, though he drew radically different conclusions from their collaboration – indeed, ones that would sit much better with Plato (Jansen 2013).

For Lippmann, the appearance of neutrality was more about reassuring readers that the public figures reported in the news have even the most urgent matters under their collective control. Lippmann believed that the simultaneous increase in political participation and purchasing power in the twentieth century made the public susceptible to mood swings that had the potential to destabilise the mechanisms of government, if not the social order more generally. In this respect, the democratic spirit was always something that 'responsible' journalism should be taming. Lippmann had witnessed how newspapers often boosted circulation by 'sensationalising' the events of the day. Such sensationalism traded in the currency of righteous indignation with an eye to revealing duplicity, conspiracy, scandal and corruption. The underlying message was: 'You deserve better than this!' Understood as a marketing pitch, it opens the door to new products that promise to satisfy the need that has been surfaced. But in politics, it can pave the way for demagogues claiming to cater to the now public's inflated expectations of good governance.

For Lippmann, the various expressions of sensationalism – not only 'war-mongering' but also 'muckraking', both early twentieth-century coinages – often resulted in poorly judged policies. Although muckraking has come to be seen positively as the precursor of today's investigative journalism, Lippmann saw it as coming from the same emotional wellspring as war-mongering. The zeal attached to both reflect a kind of wishful thinking that threatened to make the perfect the enemy of the good – and for that matter, the true. Lippmann came to an empirical understanding of this matter in the first systematic study of what we now call 'fake news', which carried the unassuming title, 'A Test of the News' (Lippmann and Merz 1920). It involved the *New York Times'* coverage of the 1917 Bolshevik Revolution in Russia, which was lauded at the time for its unprecedented immediacy and comprehensiveness, including eyewitness accounts and access to key players. Yet, once the dust settled, it became clear that the reporters' judgements had fallen victim to preconceptions and press releases. There may have even been a touch of narcissism on the part of the reporters. Many of the Russian revolutionaries – not least Lenin and Trotsky – had appropriated for their own purposes the rhetoric that the muckrakers themselves had originally used to describe poverty, corruption and injustice in America. Indeed, Lenin is reputed to have coined the phrase 'useful idiots' for the journalists that Lippmann criticised. They had been effectively 'spun'. The long-term effect was to unleash the sort of bipolar politics – part-utopian romanticism and part-existential threat – that arguably continue to characterise American attitudes to Russia to this day.

After the unexpected and – for many – unwelcomed election of Donald Trump as US President in 2016, the *New York Times* changed its motto to 'The truth is worth it', followed by four tag words: persistence, fearlessness, rigor, resolve. This combative and seemingly partisan turn would please Breitbart as it amounted to the paper's admission that it had located itself in – not above – the political fray. However, Lippmann would be disappointed. It sounds too close for comfort to the path travelled by early twentieth-century 'crusading journalism' before it was hoisted by its own petard. All that is needed now in the post-truth condition is for a third party – a Lippmann 2.0 – to establish a new baseline of truth-telling against which the *New York Times'* own biases come into open view.

For their part, politicians operating in the post-truth mode – ranging for Trump's blunt charges of 'fake news' against the *New York Times* to arch Brexiteer Michael Gove's more feline calling out of media bias at particular points in interviews – have served to level the playing field between themselves and journalists, undermining the iconic meta-level perspective of the press gallery as the 'Fourth Estate' overseeing the conduct of public affairs. Indeed, the very idea of a press gallery bespeaks a sense of privilege that post-truth's democratic sensibility finds hard to tolerate. In post-truth's level playing field, the biggest advantage that journalists enjoy is not their propensity for truth-telling but their ability to put the first spin on the data they collect. As Gove, himself a long-standing columnist for the *Times of London*, has frequently observed, one can still support a strong free press while presuming that it will always be biased vis-à-vis the stories it covers. The question that remains, even in the post-truth condition, is whether a given bias serves to improve the public's ability to understand and act on relevant matters. Regardless of how that question is answered, if the *New York Times* remains the 'newspaper of record' in the future, it will probably be due less to the accuracy of its reporting than to the brute fact that it has self-archived better and longer than its rivals.

7

SCIENCE AS THE OFFER THAT CAN'T BE REFUSED IN THE POST-TRUTH CONDITION

There are two ways to thinking about the political economy of science. On the one hand, science's internal workings tend towards what the US sociologist Robert Merton (1973) called 'communism', by which he meant that the fruits of scientific labour are collectively owned and hence freely shared. Merton was clearly talking about 'basic research' or 'pure science' as something that circulates mainly in scientific publications. On the other hand, 'science' as a socially significant activity is clearly more than simply what scientists do amongst themselves. Both public and private funders of science expect some 'return on investment' that materially benefits those outside of the scientific community. This speaks to a more 'capitalistic' approach to science. During the Cold War, the so-called linear model of science policy attempted to reconcile these two approaches to science's political economy by regarding the 'communist' and 'capitalist' ways as two successive stages of the same process. The first 'communist' phase was an incubation period during which science 'matures' so that it can be reliably applied to public and private benefit in the second 'capitalist' phase.

However, this is only a superficially satisfying solution. In particular, it leaves out the dynamic, motivational features of science. As Merton himself pointed out, science appears to be very competitive, especially in terms of scientists wanting to be recognised as 'first' in the making of some discovery. This behaviour is neither straightforwardly 'communist' nor 'capitalist'. So, some other model of political economy for science is needed. I believe that a rather expanded understanding of the 'gift' economy – one that stresses the proactive, even aggressive character of gift-giving – may help to explain

science's dynamism. After all, claims of a new scientific discovery are neither simply shared amongst scientists nor added to society's benefits. They
also force a reorientation of both the scientific community and the larger
society. Usually the reorientations are relatively minor. However, in the more
'Gestalt shift' style reorientations associated with 'scientific revolutions', they
can be seen as aggressive, as if the Mafia paid a surprise visit to your business
to 'offer' protection even though you previously thought that your business
had not been at risk. Thomas Kuhn (1970), with his winner-takes-all view
of paradigm shifts, went so far as to support what may be called an *extortion
theory of scientific change.*

Once Einstein overtook Newton in physics or Darwin overtook
Lamarck (or the Bible) in biology, 'respectable' scientists couldn't easily opt
out of the revolution and carry on as before. A cost is always incurred if
one refuses the 'new world order' that results from the revolutionary 'gift'
of a paradigm shift. In particular, one's career may be confined to the more
practical areas of the science, where the Gestalt shift matters least – at least
from a purely intellectual but perhaps also reputational standpoint. Thus,
theoretical physicists may migrate to engineering and biologists to medicine. However, even this adaptive response may not prove sufficient, once
a new cohort of scientists for whom the revolution is second nature moves
into their spaces, thereby promising less friction to the overall dominance
of the new paradigm. To be sure, if worse comes to worst, work in the history and philosophy of science is always available for those scientists on the
losing side of history.

But how exactly did we shift from science as gift to science as extortion?
Consider what distinguishes gift-giving from ordinary exchange relations.
In an ordinary exchange, the relationship between the two parties officially
ends with the completion of the transaction, in which a refusal to purchase
is not an unusual outcome, nor is it fraught with special moral significance.
However, in the case of gift-giving, the transaction is designed to start or
ratify a longer and larger relationship. Even when gift-giving isn't obviously
extortionate, the refusal to accept a gift is regarded as a prima facie normative breach, though the severity of the breach and the burden of blame are
open to interpretation. In this respect, there is a latent aggressiveness built
into gift-giving that is lacking in ordinary exchange relations. Every gift is
intended as an 'offer you can't refuse', in that you never asked for it in the first
place. In contrast, merchants are quite used to unsold goods languishing on

their shelves – and they rarely if ever blame the customers. On the contrary, 'The customer is always right'.

Here it is worth noting that what we call 'ordinary exchange relations' presupposes *a world without advertising*, which is arguably the ultimate 'gift that keeps giving'. To see what I mean, consider the world that Adam Smith imagined, which resulted in a social division of labour. One group of people conjectured that another group needed certain goods and services that the first group could provide with relative ease, or at least with comparative advantage. Thus, the second group satisfied their standing need while enabling the first group to earn a living. And if we imagine this process happening symmetrically across society, through a kind of mutual perspective-taking, then markets 'spontaneously' will emerge. To be sure, this process is fallible, since as we just suggested, markets often fail to clear – and hence goods are left to languish on the shelves. But those errors provide an incentive for merchants to stock appropriately in the future.

Gift-giving follows a different dynamic whereby the functional equivalent of 'merchants' operate much more proactively. They are the gift-givers. The strategy here is for the prospective sellers to get the prospective buyers to depend on them by demonstrating the esteem in which they hold the buyers – to such an extent that they wish no harm to come to the buyers. This initial wooing may come in the form of flattery or an outright gift. Underlying such an exchange is that the merchant knows the interests of the target audience better than the audience itself because the audience doesn't realise the extent to which its own conditions are really under the merchant's control. In this respect, merchants are in the business of converting an asymmetry into the new symmetry whereby the prospective buyers come to expect to be sold a certain bill of goods or services, which in turn means that the merchants are held accountable. A mark of this transition is 'brand loyalty' whereby consumers stick with not only a product that has proven effective but also all its future affiliated products. In our day, the so-called Apple community is perhaps the exemplar.

In high capitalism, advertising functions as the (not so) discreet visit from the emissary of the local mafia boss who makes the offer you can't refuse. In the founding text of public relations, *Propaganda*, Edward Bernays (1928) struck a suitably solicitous pose, arguing that the need for his field arose from the complexities of modern democratic society, one in which people are faced with a variety of choices on matters over which they need guidance. To the rescue comes Madison Avenue. It proceeds by reminding you of what you have always been wanting in some vague sense and then presenting the latest product as the vehicle for the satisfaction of that long-standing desire. In this respect, advertising smooths the transition from the old to the new in a rapidly changing world. In another respect, however, advertising commandeers the consumer's unconscious in a way that recalls the late Immanuel Wallerstein's (1980) explanation of how 'ordinary exchange relations' arose in medieval Europe across a vast range of goods and services through acts of piracy whereby third parties intervened to channel the flow of trade by exploiting vulnerability in the normal transport networks. In this sense, advertising succeeds by pirating the mind, or our 'neu(t)ral networks'. (I'm revisiting in a new key Harold Innis's [1951] influence on Marshall McLuhan [1964], in which control of the medium becomes control of the message.)

What connects gift-giving to advertising – and piracy, for that matter – is its high-risk character which is associated with a *proactionary* attitude to the world (Fuller and Lipinska 2014). All of these cases involve considerable expense upfront – an 'investment', if you will – in the hope of reciprocation. The strategy is to shift the moral burden to the receiving party because the initiating party has already taken a risk. Thus, the failed gift-giver is potentially humiliated, whereas the failed gift-receiver is potentially ungrateful. In this respect, gift-giving is in the business of generating *moral hazard* whereby a second party is exposed to risk simply because the first party decided to take a risk.

Thus is the logic of advertising which has occupied an increasing proportion of corporate budgets as capitalism has become more productive. More money is spent on trying to get people to buy goods than on making the goods themselves. The fortunes of companies nowadays rise or fall based not on the quality of their products but on the success of their marketing campaigns. Advances in the mining and analysis of computer user-generated data in recent years have enabled advertisers to enhance their modus operandi. Thus, in order to access a desired website, users must routinely refuse the 'gift' of customised advertising (i.e. offer of new products based on previous purchases). The result is a world that presents enormous opportunities but also a demand on our will to resist most of them. Of course, our failure to resist secures the gift-giver's income and power.

However, the presence of gift-giving in capitalism runs deeper than the advance of advertising over the past century. Nevertheless, the classical political economists failed to register it properly. They stressed efficiency as the means by which capitalism served as a vehicle for what Smith, Malthus and Ricardo called 'prosperity'. But they did not associate prosperity with endless growth. It was simply about people realising their full potential, which would in turn benefit everyone. This could be achieved in a world of scarce resources, as especially Malthus and Ricardo argued in rather different ways (Sassower 2017). For these early capitalist theorists, the greatest source of poverty was the underutilisation of human labour. People normally had neither the incentive nor the information to test the limits of their capacities by trying to do the most good with all that they had to offer. The main obstacle was an inheritance-based legal system that restricted labour mobility significantly.

Against this backdrop, language – especially its technological extension as written contracts – was touted as injecting the relevant level of efficiency into the process. What economists call the 'price mechanism' epitomises the efficiency of written contracts by reducing many potential transactions to a publicly advertised number that functions as a hypothesised transaction between the seller and any prospective buyer. Depending on whether the merchant clears his or her stock, the price will turn out to have been the right or wrong one. It amounted to a radically social constructivist conception of value – and was recognised as such in its day, not least by Kant. However, as already suggested, one radical feature of capitalism that the classical political economists failed to theorise was *its capacity to stimulate desire*, the source of the system's 'endless growth' imaginary. It has been in this context that gift-giving has played a crucial role.

Its origins lie in entrepreneurs advertising for people to work in their start-up factories. The gift in this case is gainful employment for the worker even though the exact shape of the market for the manufactured goods has yet to be determined. Indeed, there may turn out to be no such market in which case the workers still get paid even though the entrepreneur goes bankrupt. Austrian finance minister and Austrian economics founder Eugen Böhm-Bawerk (1898) used this point to launch a liberal critique of Marx's theory of 'surplus value'. Marx had famously argued that capitalist profits should not be exclusively retained by the capitalist but shared with the workers, without whom the products would not have been made. In response, Böhm-Bawerk said that Marx misunderstood the temporality of capitalism, since it was the entrepreneur who laid his or her own money, reputation and the like on the line initially to set up the enterprise that produced the jobs in the first place. If anything, the workers should be grateful, especially if they might otherwise not have been employed.

In this respect, the 'gift-giving' element of capitalist investment should be seen as continuous with the equally capitalist practice of philanthropy, which includes the charitable foundations (Rockefeller, Carnegie, Ford, Sloan, Gates, Volkswagen) that have been set up for the promotion of science. The resulting 'output' – be it from the factory or the laboratory – is seen as reciprocating the gift, regardless of the size of its actual benefit, either to the 'gift-giver' or society at large. Of course, the 'gift' here is not completely unconditional, since it may be altered or stopped altogether depending on a wide range of circumstances which normally extend beyond the original gift-giving context. But there is a general expectation that recipients of the gift will be sufficiently productive to be themselves gift-givers in the future – what is sometimes called a 'pay it forward' ethic.

The distinguished economics writer and former Rockefeller protégé, George Gilder, managed to read this gift-giving strain of capitalism back into its entire history to present the whole economic system as the rational outworking of altruism. Gilder's (1981) youthful bestseller *Wealth and Poverty* is often taken to have laid the intellectual foundation of 'Reaganomics'. Fast forward forty years, perhaps unsurprisingly, Gilder is seen today as a 'thought leader' on the broader implications of Silicon Valley, which tries very hard – albeit by no means successfully – to present its various initiatives as 'altruistic'.

Science presents an especially strong yet non-violent form of *competitive altruism*, a concept that has been used to explain the evolutionary origin of communication systems (Fuller 2006a: ch. 10). The basic idea is that to reveal something that only you know, which nevertheless may be relevant to others, is effectively to remove your relative epistemic advantage to them. In this context, the production of knowledge is tantamount to its publicity: by producing knowledge, you redistribute power. That's how communication begins, according to what the late Israeli ornithologist Amotz Zahavi (1997) interestingly dubbed the 'handicap principle'. Once the others realise the benefit of your initial signal, they will imitate your signal in some modified fashion which amounts to reciprocation. (Here one need only imagine basic behavioural reinforcement: Copy what works.) As this process scales up across space and time, a vast network of organisms and even entire species become involved in an elaborate pattern of communication that serves to sustain an entire ecosystem whereby power accrues to those who are more 'informative' in the relevant systemic sense: you become the one whose knowledge everyone else wants to have. In information-theoretic terms, what you say constitutes the difference that makes the difference.

The human practice of which all this is most reminiscent is *potlatch*, as Franz Boas (1895) famously described in his depiction of the Kwakiutl in late nineteenth-century British Columbia. Potlatch works by a chief turning over much of his wealth to a guest, who in turn is expected to reciprocate either immediately or in the future; otherwise both lose face. Boas and others have stressed the extent to which this practice undermines capitalism's reified conception of property – what Marx called 'commodity' whereby a good's value is tied to its market price rather than its social function. However, less emphasis has been placed on the character of the social relations that potlatch potentially maintains. We might envisage three types of recipient response to potlatch's aggressive gift-giving: (1) The recipient reciprocates in kind by turning over his own property to the chief; (2) the recipient reciprocates simply by returning the original gift or something quite like it to the chief; (3) the recipient fails to reciprocate at all. Whereas (1) characterises social relations at their most elite and dynamic, (2) characterises them in a relatively stable hierarchical state, and (3) in a state of disintegration.

There are analogues to these three conditions in academic citation practices. The first corresponds to the 'cutting edge' state of a field in which one publication triggers others to try to overturn its findings, either by outright falsifying them or superseding them with an even more striking finding. Histories of science are mostly about such matters. It amounts to a desire to acquire the debt carried by your predecessors, aka 'standing on the shoulders of giants'. In the philosophy of science, this idea has been expressed in terms of later theories incorporating earlier ones. The second corresponds to a normal state of play in an academic field whereby one cites those whom others cite and while trying to garner sufficient citations for one's own publications to remain competitive. Peer review is instrumental in maintaining this state of equilibrium. The third corresponds to a state of disciplinary entropy such that people are effectively ignoring each other's work: they turn down the publication 'gifts' of their peers, and they have no interest in publishing for the sake of others. The 'lone scholar' tradition of the humanities would seem to correspond to this state of mutual neglect, but in fact the vast majority of academic work across all the disciplines is routinely ignored (Fuller 2018: ch. 4).

Clearly what I have been describing here is a *mutual protection racket* which works only as long as everyone is oriented and contributing to it. In the case of the Kwakiutl, the Mafia and academia, this is made possible by the exercise of tight control over the channels of communication and reward in the respective social groups. In Boas's day, the Kwakiutl often practised potlatch in secret to avoid the notice of the colonial authorities who wanted to outlaw it as an obstacle to assimilation. Indeed, this dynamic served to turn the ritual into their primary identity marker. Of course, the Mafia's distinct brand of secret services has historically flourished against the backdrop of unreliable and corrupt law enforcement agencies (Gambetta 1988). In the case of academia, 'secrecy' comes in the form of credentials, jargon and the relatively esoteric gatekeeping practices of peer review, all of which – exacerbated by high publishers' prices – impose high entry costs on those wanting to access its knowledge. Add to that the state's role in legitimising academic authority, and the result is an epistemic hierarchy that to outsiders (aka liberal economists) can easily look like a bottleneck in the spontaneous flow of information. In this context, the internet appears as an alternative channel for publication that loosens constraints across the board – from financial to linguistic – to break academia's monopoly hold on knowledge production (Fuller 2019a).

8

WILL EXPERTISE SURVIVE THE
POST-TRUTH CONDITION?

The post-truth problem with expertise is rooted in what might be called the Standard Model of Progress. For the founders of biology (Jean-Baptiste Lamarck), economics (Adam Smith) and sociology (Emile Durkheim), a 'division of labour' in the natural or social worlds is the mark of heightened intelligence in response to environmental challenges. Just as more 'complex' – that is, functionally differentiated – organisms are more highly 'evolved', the mark of progress in the human realm is the complexity of society generated by personal specialisation (i.e. the acquisition of expertise as part of one's self-definition) and delegation of authority to an array of other equally specialised persons, all of whom are bound together in relations of mutual trust. Of course, a less flattering way to describe this situation is that both the natural and social orders become giant mutual protection rackets – perhaps even pyramid schemes – that ultimately prove only as strong as their weakest links. This would certainly help to explain why complex orders always seem to collapse in the long term (Tainter 1988).

Moreover, the Standard Model of Progress is just that – a model. There is always a prima facie mismatch between what an expert knows and the knowledge needed to get something done, especially in the sort of complex situations where the need for expertise tends to be most keenly felt. 'Expert' in today's sense is a late nineteenth-century coinage connected to 'expert witness' in legal contexts. Etymologically, it is a contraction of 'experienced'. In this regard, experts are mainly in the business of generalising from their experience as widely as possible. If you're expert at hammering, not only are you bound to look for all the nails but more importantly everything starts to look like a nail. Indeed, this tendency may even be a human version of

61

what Nick Bostrom (2014) has called 'superintelligence' in the context of advanced computers unwittingly overtaking humanity by imposing their narrow frame of reference on the world. But, to the expert's potential client base, the problem is more prosaic and deceptive. Expert advice can come across as a high-minded version of 'bait and switch' whereby the original expertise is effectively debased as experts oversell their relevance to a trusting public (Fuller 1988: ch. 12).

To be sure, this problem had been seen early. The ancestral home of expertise – the medieval guilds – had already included a moral appraisal of journeymen before licensing them to be self-employed 'master craftsmen'. One question loomed large: Does the candidate practice the necessary self-restraint to avoid exploiting potential clients by promising what he cannot deliver? If experts warn lay people to know the limits of their knowledge, then experts should be held to a reciprocal standard. This is easier said than done. Nevertheless, the principle is just as much about protecting the expert from reprisal by a disgruntled client as protecting the client from harm by the expert.

A traditional mechanism for keeping expertise in check has been the 'collegial' or 'peer review' system which is designed to generate a collective sense of ownership backed by a common standard of judgement. However, this self-policing activity is subject to the same fundamental epistemic problems that lay people face when dealing with experts. These are neatly captured in the science historian Steven Shapin's (1994) characterisation of the rhetoric of the research paper as 'virtual witnessing', a skill whose mastery and deployment may be independent of any corresponding empirical work. Indeed, arguably the true masters of virtual witnessing are the research fraudsters who have failed to get caught – even to this day. But even the ones who do get caught, especially if they managed to publish in top-tier international journals like *Science* and *Nature*, deserve honourable mention (Fuller 2006b: 105–6). I shall have more to say about this later, under the rubric of 'research ethics'.

All of this goes to show that experts fool each other as easily as they fool the public. Moreover, this 'fooling' may not even be deliberate, as experts are subject to the same cognitive biases and limitations as lay people – a finding that was first made more than sixty years ago when Paul Meehl (1954) demonstrated that his fellow clinical psychologists routinely fell short of statistically based diagnostic procedures. In that case, is there any particular epistemic advantage to the cultivation of expertise? That question migrated from philosophy to technology forty years ago, with the advent of 'expert systems', computer programmes constructed from interviews with experts that mapped out 'decision trees' for various scenarios in which their expertise might bear on reaching a successful resolution (Fuller 1993: ch. 3).

In short, the Achilles heel of expertise is *epistemic overreach*, a concept that extends well beyond fraudulently conducted and deceptively presented research, both of which involve intent. Much worse is the unintentional way in which experts support each other's excesses to minimise the size in the gaps of knowledge between them. Thus, I trust you more because I depend on you to validate my trustworthiness than because you yourself are actually trustworthy. Once the general public understands the logic of this situation, which is that of a mutual protection racket, trust can easily erode as an epistemic value altogether, resulting in the post-truth condition. Whether the latter-day 'algorithmic' descendants of expert systems, with their increased data-processing capacity and programming complexity, recover trust or simply open up new problems concerning programmer bias remains an open question, to say the least (Pasquale 2016).

However, all of the above fail to address what is often touted as the overriding value of expertise, namely, that an expert offers a relatively informed judgement on matters that the lay person may not have thought seriously about before. So, the question is really whether lay people should be able to match expert judgement against their own point of view. An interesting angle on this question is provided by public opinion research, which has been used to market ideas, products and candidates since the 1920s. Its overall effect has been to train lay people to develop 'attitudes' towards matters on which they might have otherwise defaulted to experts. A case in point is the multiple choice questionnaire which forces subjects to clarify their own thinking about matters on which they may never have focussed their minds. Indeed, Edward Bernays (1928) held that his own field of 'public relations' contributed to this smartening up of the populace – 'democratic education' – in an era when mass media was opening up new opportunities for people to exercise choice, both as voters and consumers. Nearly a century later, we should see the internet as only having increased those opportunities.

Bernays originally hoped that the brave new world opened up by market research and public relations would end the days when ignorance drove people's response to expert judgement, be it unqualified deference or out-right dismissal. Instead Bernays envisaged that people would seek *epistemic proxies* which are cheaper substitutes – or 'functional equivalents' – for knowledge that deliver the same effect in terms of satisfying their cognitive and practical needs. And insofar as such proxies foster a sense of mastery in the user, they motivate greater participations, either as consumer or voter. Even defenders of the prerogatives of expertise realise that proxies can sometimes solve the optimisation problem of acquiring the sufficient knowledge at the lowest cost in the face of an impending decision (Goodin and Spiekermann 2018: ch. 12). Indeed, when the proxies deliver an even greater effect than the originals could, they enter what Jean Baudrillard (1983) dubbed the realm of the 'simulacrum' and the 'hyperreal'. This prospect that was expedited by the rapid shift in the advertising in the 1920s from the textual and graphic to the pictorial and filmic testifies to what we would now recognise as a drive to 'virtual reality'.

Epistemic proxies are the expected products of knowledge under market conditions – at least according to neoclassical economists and their Austrian cousins (Hayek 1945; Mirowski and Nik-Khah 2017). From their standpoint, 'knowledge' is simply the name for 'authorised' information which they understand pejoratively through the histories of feudalism and mercantilism, the two economic systems that capitalism was designed to overturn (Fuller 2019a). The post-truth condition universalises this way of thinking. Nowadays the main obstacle to the promotion of epistemic proxies is *academic rentiership*, namely, barriers that academics place to the acquisition and development of knowledge, typically in the form of 'credentials', for which specific training is required before credibility is granted (Fuller 2018: ch. 4).

While seeming benign, academic rentiership may pose a much more systematic impediment to knowledge flow in society than even legally enforced intellectual property regimes. To be sure, when wearing their best face, academic rentiers promote the equivalent of 'brand loyalty', unionised 'closed shops' or perhaps 'fair trade' products. Thus, potential knowledge consumers are led to believe that there is something morally corrupt about not being informed by a certifiably credible source. The idea is that consumers are letting their desire for immediate gratification override their concern for the integrity of the process that brings it about. But is this strategy likely to prove effective in the long term, especially as people come to realise that sufficiently similar outcomes can indeed be realised by a wide variety of means? Is this concern for 'quality control' no more than mystification by another name?

The search for epistemic proxies has been arguably the modus operandi of the history of technology vis-à-vis human labour. One innovator can replace an entire class of so-called experts by arriving at a more 'efficient' means – both in terms of reliability and affordability – to solve the problems normally posed to experts. I say 'so-called experts' to recall the origins of the first 'efficiency expert', Frederick Winslow Taylor, who developed his principles of 'scientific management' in the early twentieth century in response to factory owners who were sceptical of workers' claims about the time it takes to perform certain forms of manual labour. Nowadays such scepticism is expressed about the need to matriculate at university to acquire the relevant sense of epistemic authority. Yet, as we have seen in the case of 'expert systems' and their algorithmically enhanced descendants, an over-reliance on computer technology may amount to jumping out of the frying pan and into the fire. In the end, Kant may have been right after all, that true enlightenment involves learning to trust one's own judgement.

Of course, there are ways to generate epistemic proxies by playing academics against each other, thereby attenuating the overall hold of academic power over access to knowledge. These include staging debates between acknowledged experts who hold opposing views on some issue of public concern. Academics tend to approach such events with suspicion because of the prospect that too much disagreement in open display might undermine – at least in the public's eyes – the knowledge base of their expertise altogether. This is a legitimate worry since, from Plato's Academy to the medieval guilds onwards, expert knowledge has always been shrouded in 'mystery', in the sense of its being available in the first instance only to the adepts and then later in some filtered form to the masses. This sense of mystery is enhanced by the amount of time required to access the relevant academic credentials, not least the jargon that constitutes much of the public expression of that knowledge. And so, duelling experts on stage could well turn out to be an exercise in two magicians demystifying each other's tricks, as often happens when courtroom rivals are allowed to introduce their own expert witnesses. But all of this is exactly what one should expect in the post-truth condition, where knowledge is a power game.

However, a market for epistemic proxies may be carved out of the institutionalised knowledge landscape with relatively little contestation, as in the case of non-native language mastery. It can happen in three ways. One is by learning the grammar, enabling one to read, write and speak so as to incur few errors with natives of the language. Another delves into the literature of the language to learn the style of thought and expression that the language affords, perhaps in ways that other languages do not. And the third way resists the direction of travel of the first two ways by immersing oneself in the foreign language, speaking it haphazardly and backfilling the grammar on a 'need to know' basis. This last, rather bottom-up approach sums up Max Berlitz's revolutionary technique for language learning that from its 1878 origins in Providence, Rhode Island, has become a multimillion-dollar industry adopted – and adapted by rivals – the world over. If we extend the comparison from language learning to knowledge acquisition more generally, the first method is comparable to training at a good high school or acquiring a first degree at university, the second to acquiring a doctoral-level education and the third method to acquiring the relevant knowledge through a customised series of internet-based searches. This last method is the route of epistemic proxies.

An arguably still simpler way of generating epistemic proxies for expertise is by 'going meta'. The idea here is that when faced with an expert claim to knowledge, one should assess not the claim's content (which may be too technical to be assessed first-hand) but characteristics of the field on which the expertise informing the claim is based. Even if an expert's judgement correctly captures one's field's consensus at a certain moment, depending on the frequency with which the field tends to change its collective mind, there might well be meta-level reasons for questioning the expert's judgement now. Such meta-level questioning may be apposite in climate change policy where the computer simulations on which the 'scientific consensus' is based are themselves subject to continual revision in light of new evidence and better modelling techniques. Of course, this is not to counsel 'denialism' with regard to global warming. However, 'going meta' can provide some leverage to lay people in a public sphere that regularly faces the threat of expert colonisation.

'Going meta' can be understood as a reflexively applied version of the 'second opinion' or 'independent corroboration' that professionals often encourage their clients to seek. Moreover, these modes of lay epistemic leverage are not without their own technological enhancements. Here Google's search engine and its associated 'analytics' extend the prospects for offering the lay person alternative perspectives to an expert orthodoxy. And while a widely discussed heterodox opinion is by no means a reflection of its scientific validity, the detail with which such an opinion can be developed and scrutinised does give it a vividness and concreteness that may inspire others to conduct the relevant research and tests to make the opinion a real scientific contender. In any case, the richer epistemic environment afforded by the internet has emboldened people to reclaim much of the ground that in the past had been ceded to experts. The result has been a revival of 'New Age' approaches to medicine and such scientific heresies as creationism. The implied link here between mass publication and interpretative license makes ours an age of *Protscience*, short for 'Protestant Science'. Like the original Protestant reformers of Christianity, today's Protscientists are willing to take personal responsibility – and even risk their lives – in the name of scientific beliefs that they have made fully their own (Fuller 2010: ch. 4). The rise of Protscience raises an important problem about the future of liberal democracy that can be expressed in terms of contrasting cases of dissent from the scientific orthodoxy.

First, consider the significant number of general medical practitioners and even specialised biology researchers who hold broadly 'creationist' views about the origins of life and natural history but who do not work substantively in evolutionary theory. The judgements that they deliver in their expert settings probably do not deviate substantially from what their more 'Darwinist' colleagues might say. Not surprisingly, then, the search of creationists has always resembled a witch hunt: they can easily 'hide' behind their everyday practice. Creationists and Darwinists do science in the same way, except when it comes to explaining why and how everything we know about life hangs together. And while this is a question of enormous theoretical and even existential significance, it is unlikely to affect the delivery of basic medical research and services.

In contrast, consider the case of anyone – especially a medical doctor – who opposes vaccinations for whatever reason. Such a person is potentially affecting the lives of others in perhaps an unintended yet predictable way, given the proven role of vaccination in developing 'herd immunity' to various communicable diseases. But of course, as the provenance of the phrase 'herd immunity' suggests, a policy of keeping an entire population alive has not always involved trying to keep every individual alive. Indeed, responses to the 2020 COVID-19 pandemic that focus on 'long term economic harm' from adopting a too precautionary posture to the virus – such as extended lockdowns – would seem to extend the Protscience approach even to this context.

In any case, the difference in societal consequences for the sort of dissent from expertise that characterises anti-Darwinism and anti-vaccinationism promises to put science centre stage in any future division of private and public spheres in the post-truth world: that is, issues that can be left to individual conscience versus those that require collective agreement. Indeed, a measure of a progressive liberal democracy is its tolerance of the range of beliefs to which people are entitled. Admittedly, this is a tall order in an increasingly interconnected world where the consequences of the decisions that people take based on their beliefs interact more freely, potentially raising the level of moral hazard in society, if, say, those who refuse vaccinations reach a critical threshold. In an ideal world, people can believe what they want, even if they are alone in believing it, as long as others are not prevented from believing as they wish. I have previously discussed this matter in the context of *customised science* (Fuller 2018: ch. 5). Here 'science' should not be understood as a finished product bought off the shelf, as the experts would have it. Rather, it is raw material that customers can tailor to their needs. Epistemologically speaking, you should know what the scientific orthodoxy believes without necessarily believing it yourself. In the post-truth condition, science is bought wholesale, but for purposes of retail not necessarily consumption.

In this context, experts function to domesticate the more impressionable and divisive features of democracy, which might otherwise result in the amplification of dissent. Their modus operandi sows the seeds of self-doubt, so that people stop trusting their own judgement and instead defer to 'those who know better'. In effect, experts manage to convert knowledge from a vehicle of personal empowerment to the great self-inhibitor. Yet when Socrates identified wisdom with knowledge of what one does not know, he did not mean to be endorsing rule by experts! On the contrary, he was trying to level the epistemic playing field. Nevertheless, his words have been turned against him in just this way, allowing the man behind the Socratic mask – Plato – the last laugh, as the ensuing clarity in the difference between the class of 'knowers' and the 'unknowers' has become the social structure of 'rational' democracies in the modern era, sometimes flying under the ideological banner of 'positivism' (Fuller 2006b: ch. 4). In this context, one of the most astute observers of the human condition in the modern era, the sociologist Georg Simmel, spoke of the *tertius gaudens*, a third party who benefits from other people's miseries, typically by intensifying them.

There is no doubt that expertise is an affront to democracy insofar as it implies circumstances in which others should make decisions that would be otherwise left to oneself. Of course, parliaments and other forms of representative democracy have always contained this tension, leading radical democrats from Rousseau onwards to treat 'representative democracy' as an oxymoron. If you need someone else to speak for you, then you're not living in a true democracy. Rousseau himself concluded that democracies are optimally small and populated by those who already share a sufficiently similar orientation to the world that any dissent is relatively easy to resolve into a consensual expression of what Rousseau called the 'general will'. A world governed in such radical democratic terms would amount to the anarcho-communitarian patchwork that the philosopher of science Paul Feyerabend (1979) advocated at the height of the Cold War. He projected this alternative polity as a safeguard against scientific authoritarianism (aka 'expertise') becoming the velvet glove concealing the iron fist of 'statist' (aka 'big government') rule over heterogeneous populations. Moreover, Feyerabend saw this as the dominant tendency on *both* sides of the ongoing Cold War, notwithstanding the lip service that both the Americans and the Soviets paid to 'democracy'.

Feyerabend and his fellow-travellers envisaged an ecological payoff from their proposed decentralisation of knowledge and power. After all, what the Cold War started to call 'Big Science' requires not only big brains but also enough of them working in concert to justify the allocations of resources needed to make their dreams come true, which in turn will provide the basis for everyone else's reality. That is how we got nuclear energy and genomic medicine, which together push the boundaries of humanity's hopes and fears in our own day. Playing his mentor Karl Popper in a new key, Feyerabend held that by scaling down science's grip on the world, people would be routinely subject to lower levels of risk. This would be due not only to science having less material impact on their lives but also – and more subtly – to scientists being less inhibited to admit error, precisely because less would be at stake. It would effectively put an end to the 'too big to fail' mentality that so often has turned the relationship between knowledge and power into a fatal embrace.

9

WILL UNIVERSITIES SURVIVE THE POST-TRUTH CONDITION?

It is entirely possible that, within a generation or two, academia will come to be seen in the larger society as the Roman Catholic Church is regarded today within Christianity. The Church of Rome remains the largest and most established part of the Christian religion, back to which all the other denominations can trace their ancestry. Moreover, it continues to present itself as the 'universal' church, the original Greek meaning of 'catholic', which in turn allows the Pope to label dissenters as 'heretics' and subject them to 'excommunication' whereby they are no longer treated as Christians. Nevertheless, for the past five hundred years Catholicism's authority over matters of faith and practice has very perceptibly ebbed away for those who still call themselves 'Christians'. The main charge levied by the various lines of Christian dissent normally called 'Protestantism' against the Catholic Church is that it is too clerically mediated and hierarchical to respond to a changing world at a level that takes seriously both not only their existing needs but also their untapped potential. Thus, the main growth areas for Christianity in recent times have come from evangelical Protestants spreading the gospel in Latin America, Africa and Asia by making extensive and innovative use of the internet, updating the Protestant Reformation's original strategic deployment of the printing press as a communicative medium (Eisenstein 1979).

The analogy to contemporary academia should be already apparent from this general description, but it can be made clearer by delving into the motives of the original Protestant, Martin Luther, who was professor of Moral Theology at the University of Wittenberg, when on Halloween 1517 he posted his *Ninety-Five Theses* to his clerical superior, the Archbishop of

Brandenburg – and figuratively (i.e. nailed them) to a Wittenberg church door. At the heart of Luther's concerns with the Church to which he had devoted his life was its practice of selling *indulgences*. In today's secular academia, similar worries increasingly attach to the dispensation of *credentials*.

An indulgence is a clerically brokered price on salvation, which normally involved 'good works' that combined confessional rituals and financial payment, in return for which sins would be forgiven. An indulgence could also absolve the dead of their sins so as to hasten their passage out of purgatory to heaven. The practice did not seem especially strange to those familiar with the biblical account of Jesus appointing his chief disciple, Peter, to do God's business on earth once he was gone. Peter's papal successors further delegated God's business to an elaborate and dispersed network of church officers – bishops, priests and the like – who enjoyed considerable discretion in the dispensation of indulgences.

Before explaining Luther's outrage at this state of affairs, consider its contemporary analogue in higher education today. Those who pay to attend university have been traditionally led to believe that their degrees entitle them to lifelong employment, a secular heaven on earth. They supposedly confer a privileged epistemic state, 'credentials', which effectively convert anything that their possessor says from mere opinion to authorised knowledge. In this respect, universities offer themselves as vehicles for the public absolution of ignorance by requiring students to pass exams based on courses of study designed to get them to work off their ignorance. Thus, all the classes attended and the assignments submitted amount to 'penance' on the path to epistemic absolution.

Back to Luther. The dispensation of indulgences amounted to a perversion of the Christian gospel. For Luther, it is in the gift of God alone to forgive, a power that the deity exercises without fear or favour. The very idea that salvation might be purchased at a price as stipulated by officers of the church was an affront to God's autarchy. Moreover, Luther believed that the uncertainty of divine redemption actually served to improve humanity's moral character, regardless of whether particular individuals turn out to be saved or damned. Indeed, one's entire life should be seen as a trial of the soul, from which at the very least others might learn. In contrast, indulgences encouraged people to avoid confronting the causes of their erroneous ways, and simply carry on as they always have.

Might not something similar be said about the academic dispensation of credentials? Whether your knowledge of, say, physics is any good depends on whether it enables you to solve practical problems in the physical world, as engineers and inventors routinely do, or even to solve more theoretically based problems that constitute a genuine contribution to the foundations of science itself. These demonstrations of knowledge are different from what is on display when you pass enough physics exams to acquire a physics degree. Indeed, to evaluate job candidates by their grades and degree classes is arguably tantamount to 'virtue signalling' in its most objectionable sense. After all, to be good as a student is not necessarily to be good as a teacher, let alone a practitioner or applier of knowledge in the field one has studied. It simply shows your ability to conform to the norms of the institution dispensing the credentials – a dubious proxy for genuine knowledge. Yet, this is what today's universities would look like to Luther. In a backhanded sense, we have long already known this, which has resulted in both a 'Reformation' and a 'Counter-Reformation'.

The Reformation strategy is to deny the need for credentials in hiring and instead set up standards of knowledge appropriate to the task at hand, as dictated by the employer. One of the world's leading accountancy firms, Ernst & Young, and the UK's leading Tory magazine, the *Spectator*, have begun to administer their own in-house examinations, which are open to anyone who wishes to apply. More aggressively, the Silicon Valley venture capitalist Peter Thiel launched the 'Thiel Fellows' in 2011 whereby top-flight high school graduates are lured from elite universities to spend time developing innovations to bring to market. In these cases, the employer or funder takes full responsibility for certifying candidates, without any academic mediation.

A more profound Reformation tendency has been quietly underway for the past twenty years. It involves internet users turning the medium into an all-purpose knowledge acquisition device – if not a literal extension of their selves, especially if social media is taken into account. Indeed, the field of 'cyborg law' has emerged, reflecting a shift in personal preference from expert human mediation to a more straightforward personal identification with the technology on which one depends for information (Wittes and Chong 2014). To be sure, some human mediation is always involved. What ends up on Wikipedia is more mediated at the strictly epistemic level than what ends up on an ordinary Google search, which is also mediated, but by user practice and sponsor considerations. But no one denies that what appears on the internet is subject to much less mediation than what appears in academic publications, which are subject to notorious 'peer review' vetting procedures.

The turn to the internet is comparable to the role that printed bibles in the 'vulgar' languages of Europe played during the Reformation, which potentially enabled every literate person to receive the gospel in their own tongue. Like the internet, the Bible contains multiple perspectives. More importantly, the Bible invites multiple interpretations of those perspectives, since the human reader is addressed from the outset as having been created 'in the image and likeness of God'. It might not be too fanciful to suggest that this was the original moment of 'user-friendliness' whereby literature began to acquire the properties of self-empowerment which the internet has inherited. The original long-term result was to generate denominations of Christian and 'para-Christian' religions (e.g. Unitarianism, Christian Science, Mormonism), which coexist with the more traditionally accredited versions to this day. It is the sort of future into which we seem to be heading, in terms of the relationship between academic and non-academic knowledge.

Nevertheless, for the moment we live in the midst of a 'Counter-Reformation' whereby academia is doubling down on its epistemic authority by raising its entry costs. That people seeking regular academic employment need to acquire ever more credentials – multiple degrees and postdoctoral appointments – speaks to this point. Moreover, academia has fostered a culture of moral panic over the potential corruption of the knowledge system through a heightened awareness of research fraud, student plagiarism, as well as the persistent levels of public resistance to the scientific consensus around climate, evolution and medicine. Taken together, these rather different issues point to an academic establishment focussed more on preserving its own integrity than on promoting the growth of knowledge per se. The logic suggests that over time individuals need to invest more heavily in the credentialing process before being accredited in what is on offer. The turn taken by Ernst & Young and others implicitly questions the logic's long-term sustainability. The costs are beginning to look too high and the benefits too little. So what can academia do to ensure that it survives the Reformation happening in its midst?

Most obviously, academics should not add to student entry costs by requiring them to reproduce the process by which the academics themselves acquired knowledge of their subject. On the contrary, the teachers should endeavour to make the knowledge easier to acquire, shearing it of its scholastic trappings and presenting it as something of utmost urgency to the students' lives. In this way, the classroom experience might approximate what the Protestants achieved through 'evangelising' their faith. This is another, albeit more democratically explicit way of expressing the so-called Humboldtian mission of the modern university to present one's research as something worth teaching to people who start from a radically different place from you. Although evangelists have been in equal measure revered and mocked, their modus operandi has reflected very well an awareness that prospective converts have a range of faiths from which to choose, none of which holds an obvious monopoly on the truth. In such a market for what Stark and Bainbridge (1987) have called 'transcendental goods', one aims to get students to take the risk of exchanging their old ways of thinking and being for new ones. In the end, the benefits should accrue to students in the actual conduct of their lives and not simply their ability to keep paying into the academic system.

The bottom line is that making the university fighting fit for the post-truth condition requires significant changes to its structures and functions. These will involve a makeover – but *not* the abandonment – of the classic Humboldtian unity of teaching and research. The physical sites of the Humboldtian mission – the lecture hall and the seminar room – will need to be refashioned in the spirit of theatre and spectacle, venues whereby students are given a unique opportunity to see what it is like to embody inquiry in one's own person through speech and other rhetorical displays. Teachers will no longer spoon-feed students with knowledge that is sufficient to pass exams but leaves a bad aftertaste – the pedagogical equivalent of 'junk food'. Instead, they will encourage and enable students to make whatever they learn integral to their personal development, what Humboldt as an heir of the German Enlightenment would have recognised as *Bildung*. In practice, it means that pedagogy becomes a species of *dramaturgy*, a term fashioned by the original German Enlightenment thinker Gotthold Lessing to describe the interpretation of scripts for purposes of performance in various dramatic settings – or, in blunt public relations terms, the translation of message to medium. Thus, the excellent academic would be the excellent dramaturge, a kind of 'meta-director' in the theatre of knowledge, capable of showing to students how a particular concept or finding might apply in real-world contexts (Turner and Behrndt 2007).

The COVID-19 pandemic presents a unique opportunity in this regard, especially if universities wish to retain their current enrolment levels and yet are required to house fewer students on campus due to social distancing measures. In that case, one can imagine two types of student experience, one corresponding to that of *actors* and one to that of *audience* (which may be priced differentially, of course). The actor-students will more explicitly try out what they are learning in an extended sense of the 'classroom'. In the past, such extended campus-related activities have been often tied to student unions, increasingly under the rubric of the 'co-curriculum' (Fuller 2018: 125). These include debating societies, radio stations, external speaker events, as well as charity and activist work. But universities could be bolder and dedicate specific spaces for student re-enactments, dramatisations and experiments based on what they are learning. Continental European universities have traditionally allowed student-organised seminars and reading groups that led to petitions for academic credit and even the hiring of lecturers capable of addressing topics that fall outside the official curriculum.

One strategy would be to revisit the dramatic use of space in the original amphitheatre-like design of 'lecture theatres' and 'operating theatres' in nineteenth- and twentieth-century universities. A bit more adventurous would be to combine lecture and seminar formats in large halls with a suitably distanced cabaret-style seating arrangement that would allow the lecturer to speak in motion, like a singer. In any case, lectures are already routinely recorded at many universities but they could be made worth watching by a more general audience, in which case the students on site may function as a 'studio audience'. Indeed, if the university in its current form – with its vast holdings and high aspirations – is to survive, it should think of itself as a *drama school on steroids*. This means seriously investing in the 'production values' of audiovisual entertainment, both at the technological and the human levels. It follows that academics should be hired primarily for their ability to act out what they think students should be learning. It is worth recalling that the dramatic setting of the medieval doctoral examination whereby the candidate publicly defended the supervisor's honour against a designated 'opponent' was the original site of 'academic charisma' (Clark 2006). Arguably the 'performative' pedagogy surrounding the 'social justice' agenda continues this tradition of education as high drama – albeit in its own inimitable way.

Such an academic leap into 'hyperreality' would amount to quite a creative response to the post-truth condition. Universities could rival the mainstream entertainment industry, repaying the compliment of scientists' extensive involvement in the development of futuristic films since the 1920s (Kirby 2011). To be sure, the formal assessment of such activities remains an open question. It should probably be relatively broad and perhaps even perfunctory – that is, not propped up by artificially precise marking and gratuitously elaborate feedback that only serve to inhibit future exploratory performances. In any case, assessment should be disconnected from the dispensing of professional credentials, which is likely to become less of what universities do anyway.

10

'RESEARCH ETHICS' AS POST-TRUTH PLAYGROUND

'Research ethics' is a figment of the regulatory imagination. It conforms to neither of the bases on which codes of professional conduct have been grounded: *transactional* and *transcendental*.

A *transactional ethics* governs the parties to an exchange. The exchange between a scientist and a prospective subject in an experiment should be a case in point in which the scientist would inform the subject of all the risks involved in participating in the experiment, including whatever insurance is available in case of adverse outcomes. The prospective subject would then decide whether or not to participate. Indeed, such transactional ethics have been strongly supported by 'transhumanists', who believe that in order for humanity in the future to progress as fast and far as it already has, it needs to promote greater subject involvement in research (Fuller and Lipinska 2014: ch. 4). Yet the academic 'institutional review boards' that administer research ethics are much more paternalistic than transactional ethics would permit, as they prohibit from the outset certain sorts of experiments that are deemed too harmful to potential subjects, regardless of their personal risk thresholds. But if subjects are indeed as impressionable and vulnerable as the research ethicists insist, then perhaps we should put a halt to all empirical studies of humans, not only experiments but also ethnographies and even interviews. Luckily, at least for the sake of science, so far it hasn't come to that state of affairs.

A *transcendental ethics* aims to promote what is in the best long-term interest of the activity in question. Science would seem to be ripe for this treatment. Indeed, philosophers of science have long recognised a distinction between the 'context of discovery' and the 'context of justification', according to which it doesn't matter how you arrived at an idea or finding ('discovery') as long as it contributes to science's overall direction of travel ('justification'). This distinction is normally invoked to excuse the 'exotic' religious, metaphysical or political views of accomplished scientists – as being 'merely' part of the discovery process. But the distinction can be equally deployed to excuse scientists whose 'exotic' moral sensibility – perhaps a version of 'the end justifies the means' – allowed them to overstate if not fabricate knowledge claims upon which others nevertheless managed to build productively, thereby effectively 'justifying' their original pseudo-findings. Galileo and Gregor Mendel come to mind as obvious cases in point (Fuller 2007: ch. 5). The post-truth question is not whether such 'errors' are pervasive in all of science but the extent to which they impede or facilitate the ends of science. And that's ultimately a judgement call, one continually made over the history of science. In that case, the interesting question is who calls the judgement at any given time. Yet research ethicists want to prohibit any such 'exoticism' from ever entering science, regardless of the benefits it might bring.

The curious character of research ethics doesn't stop here. Even the core concept of *evidence* is fraught with ambiguity if not outright equivocation. For example, 'reliable' means both something that actually happens on a regular basis and something or someone that you are inclined to trust. The former doesn't necessarily result in the latter, and the latter is rarely based on the former. Yet this 'constructive ambiguity' has been central to the 'social theory of truth' in the modern era whereby one comes to associate certain facts about the world with certain people who are authorised to speak on behalf of those facts (Shapin 1994). Indeed, one contemporary philosophical account of knowledge, 'reliabilism', promotes this ambiguity into a full-throated defence of 'epistemic paternalism' (Goldman 1999). What makes the situation ambiguous is that the identities of the relevant people and facts are so intimately bound up with each other that it is difficult to tell whether one believes the people because of their 'reliable' access to the facts or the facts because of the 'reliable' people who have got access to them.

The referencing practices of academic texts demonstrate this ambiguous sense of 'reliability'. To make one's own claims appear credible, one must cite precursors who paved the way, in the dual sense of providing not only building blocks of knowledge but also moral assurance that one's own construction is based only on credible sources. But in reality, as we saw when considering science as 'gift-giving', this is just a kind of backward-facing 'pyramid scheme' whereby the reader is led to believe that behind the academic references are indeed those 'hard facts' that can serve as building blocks for the author's epistemic edifice. Yet these supposed bearers of hard facts are imposed by journal editors on those wishing to contribute to the field over which the editors hold sway as 'gatekeepers', a term that recalls the medieval origins of toll roads to constrain the passage of travellers through privately owned land. Nevertheless, in the spirit of what Nietzsche called 'genealogy', which the analytic philosopher Robert Brandom (1994) has antiseptically updated as 'anaphoric reference', we might try to retrace the stages of the academic pyramid scheme by following the trail of citations, only to discover that such 'facts' are not quite as 'hard' as one might have expected. In that case, the pyramid is reduced to a house of cards. Indeed, in fields where both scientists and journalists are highly motivated to find error because of the political or economic salience of the research, this has periodically turned out to be the case.

Moreover, this practice extends to students, around which some appalling pedagogy has developed. Instead of finding their own voice, students are instructed to prioritise looking for authorities who anticipated what they would like to say. It results in a weird kind of ventriloquism that is sometimes called 'dummy citation'. This is the practice, routinely found in both student and academic writing, of crediting 'leading figures' with discipline-based truisms in order to demonstrate one's own worthiness to contribute to the field. In both contexts, one's own contribution is needlessly minimised, while the significance of one's precursors is artificially inflated. All of this is unfortunately in line with the rent-like 'entry costs' that structure both the education and research sides of academic life (Fuller 2019a).

Of course, there is value in studying those who have previously followed a similar line of inquiry. But much of that value may be realised by effectively recycling old content in a new context. The student who cuts and pastes an earlier work in a way that satisfies the demands on an assignment – what nowadays would be classified as 'plagiarism' – is acting no differently from a poet who succeeds in obliterating the memory of those whose words he or she has recycled. Both the student and the poet have exercised critical judgement, the proof of which lies in its reception. When aestheticians say that every great artist is a great critic, this is what they mean. Great artists know what is worth using, and they use it well. The recent educational focus on 'curation' aims to recover this attitude from the academic obsession with plagiarism (Fuller 2016: 44–46). Indeed, the 'anxiety of influence' that poets routinely face is not that they might be caught for plagiarism but that if caught, they will be judged to have produced a work inferior to the original (Bloom 1973). In any case, whether plagiarism amounts to an original work or merely a poor copy of the original is a question that should be decided by the market not by self-appointed intellectual vigilantes acting in the name of 'research ethics'.

Here the law takes a more sensible view. Legally speaking, the fixation on plagiarism gets the point of assigning property rights to intellectual products exactly backwards. The point should not be to create an endless trail of debt whereby those who come later must always pay backward to their predecessors before proceeding forward (Frye 2016). On the contrary, the point of intellectual property rights is to ensure that those who come first enjoy only a temporary advantage before others appropriate the work to their own potentially greater advantage. 'Intellectual property' – defined either in terms of ideas or words – is something that could have been gener-ated by anyone, and only a matter of circumstance enabled a particular indi-vidual to come first. This principle tracks a basic intuition about how ideas and words come to have value – namely, from the contributions of many to the benefit of many. In this respect, 'intellectual property' is ultimately about collective ownership. Indeed, it was a cornerstone of Pierre-Joseph Proudhon's original nineteenth-century formulation of the philosophy he called 'anarchism' (Guichardaz 2019).

It is interesting to see what 'research ethics' looked like before the intel-lectual property frame came to dominate our understanding. As late as the final quarter of the nineteenth century this was still possible. Back then the discussion was bookended by famous lectures on the 'ethics of belief' given by the mathematician William Clifford (1999) and the psychologist William James (1960). Clifford argued that one should never believe more than what the evidence shows, whereas James argued that the inevitably partial nature of evidence means that one must always go beyond the evidence, which he described in terms of the 'will to believe'. To their original late Victorian audiences, it was clear that Clifford was trying to draw a sharp line between science and religion, and James trying to blur it. While Clifford was generally sceptical about the use of hypotheses in science, James likened hypothesis-testing with trials of faith. The one treats the necessary uncertainty of our knowledge as a threat to be avoided, the other as an opportunity to be embraced.

Nowadays, armed with the methods of statistical inference, we would say that Clifford's research ethics policy is prone to 'Type II' errors ('false negatives') and James's to 'Type I' errors ('false positives'). In the former, one fails to see something that is significant, whereas in the latter one sees something that is really not significant. In terms of science's historical track record, James would seem to have the upper hand, at least insofar as 'progress' has resulted from hypotheses that ventured far beyond the available evidence when they were first proposed but turned out to be right in the long term. More to the point, that initial faith in an idea, however vague or nebulous, has been necessary to motivate the inquiry that finally refined it into an agreed truth. Think of it as a secular update of St Augustine's aphorism: *Credo ut intelligam*. ('I believe in order to understand'.)

One factor that complicates this disagreement over the ethics of belief is that standards of 'evidence', let alone 'proof' – however one defines these terms – are not uniformly applied. Indeed, what is called the 'scientific establishment' relies on upholding a clear distinction between 'presumption' and 'burden of proof' (Fuller 1988: ch. 4; Fuller and Collier 2004: ch. 10). One must not only cite the right precursors but all too often must also come from the right place, or at least possess the right academic pedigree, simply in order to get a fair hearing. This latter point was brought into high relief in a landmark social psychology experiment in which already published articles were resubmitted to their original journals, except that the articles' authors were given fictional names and presented as coming from low rather than high prestige institutions (Peters and Ceci 1982). Most of the resubmitted articles were rejected, even though the editors had failed to detect that the articles had been previously published. This study helped to initiate the current academic practice of 'blind peer review', whereby the author's identity is masked prior to the formal assessment of his or her article's merit. Interestingly, the successful plagiarism that was required to make the experiment work in the first place was hardly remarked upon at the time.

To be sure, Ceci and Peters (2019) realise that their experiment would not pass today's institutional review boards, given the level of deception to which the subjects – in this case, journal editors and reviewers – were submitted. Nevertheless, the four decades following their research have witnessed the rise of a new genre of deceptive research practice, arguably conducted very much in a post-truth spirit. This genre draws inspiration from the so-called Sokal Hoax of 1996 whereby the physicist Alan Sokal published in *Social Text*, then one of the world's leading cultural studies journals, an elaborately referenced article that alleged to draw wide-ranging connections between cutting edge work in mathematics and physics and politically correct post-modern social theory (Sokal and Bricmont 1998). However, as soon as his article was published, Sokal revealed to various news outlets that the elaborate references amounted to an elaborate joke on the credulous editors. The story made the front page of the *New York Times*, triggering some inordinate soul-searching about whether the humanities and social sciences uphold proper academic standards, a debate that rumbles on even to this day.

I say 'inordinate' because from a post-truth standpoint, the most striking feature of the Sokal Hoax is just how much Sokal's own framing of the situation prevailed in the minds of both journalists and his academic targets. Because neither had expected the hoax, he was given a free pass to dictate the name of the game in which he positioned them as players. Thus, the *Times* uncritically published Sokal's version of events, putting the journal editors on the back foot to defend their actions. More regrettably, the editors then abandoned their own postmodernist scruples by conceding Sokal's interpretation that his text was indeed 'fashionable nonsense'. As someone who had also published a genuine(!) article in that ill-fated *Social Text* issue, I remarked on this failure of nerve both at the time and subsequently (e.g. Fuller 2006b: secs. 19–20). After all, the great postmodernist gurus Roland Barthes, Michel Foucault and Jacques Derrida had declared the author 'dead' in the sense that a text's meaning is not univocally imparted by its writer but multiply interpreted by its readers. Accordingly, the *Social Text* editors should not have been spooked by Sokal because by their own lights Sokal is no more than just another reader of his own text. Although he might regard his text as a fraud, if a sufficient number and range of readers derive meaning and benefit from what he wrote, then his own interpretation will have been proven incorrect. Needless to say, the postmodernist editors of *Social Text* failed to 'walk the talk', as Americans say.

An especially bitter aspect of the Sokal Hoax is that Sokal knew at least one of the *Social Text* editors, whose leftist politics he shared and made much of in the hoax article. This contributed to a sense of betrayal, once the hoax was revealed, which made lessons from the episode still harder to learn. Had Sokal's article been subject to blind peer review, much of that sting would have been absent. For this reason, the recent 'grievance studies' – or 'Sokal Squared' – controversy is perhaps more interesting in terms of shedding light on whatever 'ethics' might be at play in the conduct of research (Lindsay et al. 2018). Here the authors sent twenty articles to a wide range of cultural studies journals that review submissions blindly. The hoax was uncovered only when one journal had to contact a named author and discovered that she did not exist. The article in question – on 'dog rape' – was already published and its peculiar content had even attracted the attention of the *Wall Street Journal*. By that time, it was one of the four pieces that had already been published, with three accepted but not yet published and seven still under review. Only six of the original twenty submissions had been outright rejected.

While much has been predictably made of the gullibility of the editors of the various cultural studies journals, now especially given their blind submission process, much more significant is what Sokal Squared shares with the original Sokal Hoax – namely, the authors' savvy to get ahead of the story. Once the story broke, they spun it in a way that allowed them to dictate the name of the game. The Sokal Squared principals quickly released an elaborate documentation and justification of their actions, helpfully supplemented by an 'independently' crafted YouTube video, suggesting that they had realized that their ruse would be revealed at some point. Yet in displaying such public relations savvy, they effectively undermined the spirit of scientific gamesmanship as it has developed in the modern era. But the Sokal Squared authors are by no means alone in this sabotage. Indeed, they were simply acting in the same spirit as 'legitimate' scientists who reveal their 'discoveries' in press releases before they have appeared in a peer-reviewed journal, let alone before any substantial uptake by the target research community.

While those celebrated research fraudsters Galileo and Mendel would not pass today's research ethics watchdogs, they are not seen as having caused lasting damage to the fields concerned. That's because the ultimate role played by an academic publication in the knowledge system is normally determined in the *post-publication* phase – that is, by its community of readers, who decide how to use and/or extend the research in productive ways – or criticise if not avoid the research altogether. This long-term process eventually overshadows whatever benign, malicious or mischievous motives informed those who originally conducted the research. In short, the 'hoaxiness' of the project undertaken by the Sokal Squared collaborators – like that of their role model Sokal – depended on the authors' success in pre-empting this post-publication process by not allowing people a fair chance to judge for themselves the merits of their pieces. Rather than simply let the articles attract – or not, as the case may be – a community of readers, the authors immediately declared what they had done: Because they planned their pieces as part of a hoax, you should therefore understand what they have done as a hoax.

But there is no reason why the audience must succumb to the spin that either Sokal or the Sokal Squared collaborators have given to their work. For example, the Sokal Squared collaborators wrote their articles partly with the help of a 'postmodern word generator'. What they did is really no different from artificial intelligence researchers who try to design algorithms capable of converting data into publishable scientific articles. In both cases, blind peer review functions as a kind of 'Turing Test' for the would-be algorithmic author. Indeed, a thoughtful defender of the Sokal Squared hoax admitted that he thought that at least one of the articles was quite interesting in its own terms – that is, before he learned that it was the product of a 'hoax' (Soave 2019). Moreover, if the philosophical distinction between the contexts of 'discovery' and 'justification' in science obtains, then why should the intent to deceive matter at all in the evaluation of a piece of research? After all, peer review – *especially* when it is blind – is meant to test not the author's moral character but only the possible epistemic value of his or her work for the target audience. Indeed, to think that the integrity of science requires the integrity of individual scientists is to commit what logicians call the 'fallacy of division'. The history of science is full of cheats, liars and hypocrites who pushed forward the frontiers of organised inquiry – and most of them may be alive now. In the end, research ethics is simply a species of risk management, where the *de facto* – and desirable – policy in the history of science has approximated William James's 'will to believe'.

11

WHY IGNORANCE – NOT KNOWLEDGE – IS THE KEY TO JUSTICE IN THE POST-TRUTH CONDITION

Perhaps the most consequential yet under-reported dispute in political theory over the past half-century has been whether *strategic ignorance* or *comprehensive knowledge* is the key to justice. Self-described political 'progressives' have been on both sides of the issue. But their actual political affiliations have been 'nomadic', to say the least. To see what I mean, I shall cast John Rawls, upholder of the left-liberal establishment in the 1970s, as the patron saint of the ignorance-based 'post-truth' position on justice, whereas Robert Nozick, his younger Harvard colleague and main right-liberal opponent at the time, turns out to underwrite the knowledge-based 'truth' position associated with today's 'politically correct' ideas of restorative justice. What has remained constant is that the 'ignorance' side sees justice as residing in the future, whereas the 'knowledge' side sees it as residing in the past – in short, *risk management* vs. *score settling*. The one veers towards an imaginary utopia on which we should train our sights, the latter towards an original trauma that requires remediation.

On the side of ignorance was the greatest philosopher of the social democratic welfare state, John Rawls (1971). He argued that in order to decide the constitution of the just society we must be in a state of strategic ignorance, which he likened to a veil. To be sure, Rawls would allow all knowledge of human societies past and present to enter into our deliberations. Yet we must act as if we would not know our exact identity in this hypothetical society. Once the veil is lifted, we could turn out to be of any race, gender, class and so on. Rawls's elaborate thought experiment aimed to capture a core intuition underwriting the concept of justice, namely, *fairness*. For Rawls, the just society is organised as a fair game. Although Rawls's defenders would be loath to admit it, that is a very post-truth way of seeing things. Accordingly, some forms of ignorance trump knowledge for strategic reasons, perhaps even resulting in a higher form of knowledge, as justice has been often characterised from Plato onwards. Rawls's 'veil of ignorance' – symbolically playing on the early modern personification of justice as a blindfolded woman – is designed to establish what amounts to a 'level playing field' in judgement.

On the side of knowledge was perhaps the most philosophically sophisticated defender of modern libertarianism, Robert Nozick (1974). He tied justice to what he called 'entitlement', which implied knowledge of the history by which people acquired their current stakes in life, including both their capacities and their liabilities. To be sure, Nozick simply took one's native capacities and liabilities as given, even though one could query how one's starting point in life had been 'genetically acquired', to give a libertarian spin on inheritance. Nevertheless, Nozick was mainly interested in the terms under which transactions involving significant changes of identity occurred in one's own lifetime: Was a transaction free or coerced? Were the parties sufficiently informed about the nature of the exchange? Unlike Rawls, for whom it was paramount that the just society enables all of its members to lead a good life, Nozick was more relaxed about people making bad decisions that indefinitely disadvantage them. His main concern was whether such decisions had been taken freely and knowledgeably. Very often they have not been: the original balance of power and knowledge has been so asymmetrical that one side was effectively coerced and deceived. For Nozick that constituted injustice.

A theologian would naturally see the difference between Rawls's and Nozick's approaches to justice as alternative secular versions of *theodicy*, the justification of God's ways to humanity. Rawls aims to maximise salvation for everyone, regardless of their individual capacities and liabilities in life, whereas Nozick wants to put everyone in the place of Adam before the Fall, on the basis of which they acquire their individual capacities and liabilities in life. What specifically matters for our purposes is Nozick's need to establish specific factual truths about the conditions of the original transaction before setting the basis for policy today.

Both Rawls and Nozick were adopted by groups who considered themselves politically progressive. Rawls helped to justify 'race-blind', 'gender-blind', 'class-blind' policies in university admissions and employment decisions. To be sure, this sometimes took the form of 'affirmative action', the practice of tipping the balance in favour of historically disadvantaged groups when all other qualifications are equal. However, this was applied in the spirit of handicapping a horse race with the ultimate aim of achieving a level playing field. In other words, the long-term success of any affirmative action policy would be measured by its ultimate redundancy (Fuller 2000a: ch. 4). The overall focus was on the sort of society in which we and our descendants would want to live – not on redressing historical injustices as such. Indeed, positive value was generally placed on 'breaking with the past', if not quite forgetting it, which provided an 'institutionalist' correlate to the 'radical' and 'revolutionary' self-understanding of the 1960s and 1970s. Although technocratic social democrats have been the staunchest Rawlsians in recent times, a renovated Rawls is also suitable for the post-truth condition, one which takes seriously that our powers of thought are enhanced by the sort of future-oriented strategic ignorance on which the Rawlsian veil trades.

But Nozick equally attracted strange bedfellows. He became the natural ally of those who claimed that their ancestral property had been acquired without proper consent – indeed, typically by violating the owners themselves. In Nozick's day, this charge was most naturally heard from indigenous 'Indian' Americans. However, Nozick-style arguments for what is nowadays called 'restorative' or 'reparative' justice are increasingly mounted on the back of the generally successful 'Truth and Reconciliation' committees that were set up in the wake of various politically inspired killings in Latin America and Eastern Europe from the 1970s and 1990s, and most notably in post-Apartheid South Africa. Self-styled 'postcolonial' academics and lawyers in the US have extended this line of argument on behalf of the descendants of Africans who were abducted into the Atlantic slave trade from the sixteenth to the nineteenth centuries. US journalist Ta-Nehisi Coates (2014) has eloquently defended the position by detailing the continuing downstream effects of slavery, notwithstanding 150 years of Constitutional abolition and more than fifty years after the passage of the Civil Rights Act. Although this radical argument for reparations has so far gained little traction in the courts, it has succeeded in questioning the wisdom of the original UK-inspired US policy of 'compensated emancipation' whereby the federal government paid out 'reparations' to former slave owners – not the slaves themselves – in the 'Reconstruction' period following the Civil War.

Postcolonial restorationists share Nozick's suspicions about the strategic deployment of ignorance as an instrument of justice. For them the appeal to ignorance is a barely concealed power move (McGoey 2019). What Rawls took to be a liberating feature of his thought experiment, Nozick had already found potentially oppressive, as the 'veil of ignorance' literally requires that people repress aspects of their own histories to facilitate acceptance of Rawls's welfarist utopia. Today it is easy to forget that Rawls's approach was intuitively accepted by most academics and policymakers when it was first presented, which made Nozick's complaint seem like sour grapes. But today the tables are turned, as Nozick's unlikely heirs wreak revenge on Rawls's strong activist state (e.g. Forrester 2019). They speak in the name of 'epistemic injustice' and their followers are often pejoratively called 'social justice warriors' (Fricker 2007 ; Collins 2017). In these warriors' line of fire are not only policymakers but also – and perhaps especially – scientists, who are accused of suppressing relevant knowledge both about and from historically disadvantaged groups in order to prop up a vision of the world that justifies the current 'straight white male hegemony'. This in turn gives the lie to any 'universality' that the policymakers and scientists might claim on behalf of their views.

The ultimate foundation for all these concerns is Nozick's philosophical polestar, John Locke, who argued that consent is required of all those who would be covered – and hence governed – by any form of knowledge or power that claims to be universal. We owe regular democratic elections to this Lockean insight whereby 'the people' are given the opportunity to offer a new mandate to the ruling party or transfer power to another party that represents the people's new collective self-understanding. Interestingly, no procedure remotely comparable exists for how scientists reach agreement among themselves, let alone with the larger society, concerning knowledge claims of potentially universal import. To be sure, science-based controversies are frequently punctuated by rhetorical appeals to that sociological mirage known as the 'scientific consensus'. Whatever this 'consensus' might be, it has nothing to do with democratic elections, even among the scientists themselves. Rather, it refers mainly to the science's 'self-organising' gatekeeping processes. As we have already seen, these tend to reinforce science's own status hierarchies – along with the more encompassing social hierarchies involving race, class and gender that typically trouble social justice warriors (Fuller 2018: ch. 3).

The current political salience of restorative justice is largely due to the South African case in which the historical injustices legitimised by Apartheid legislation were still in effect until 1991, after which South African citizen rights were no longer circumscribed on racial grounds. However, in most cases of injustice for which reparations are claimed, the continuity between 'cause' and 'effect' is not so straightforward. A certain amount of time will have passed – perhaps even centuries – between the alleged commission of the injustice and the harms allegedly still felt. This can make adjudication fiendishly difficult, even for those sympathetic to the restorative justice agenda (Elster 2004: ch. 3). Aside from the sheer disappearance of evidence, there is the no small matter of identifying 'perpetrators' and 'victims' in terms of the current population, given that the original individuals are likely to be deceased and their descendants have moved in all sorts of ways involving the mixture of capital, genes and culture, resulting in identities whose legitimacy does not trade on the distant past (Cowen 2006).

The Atlantic slave trade is a good case in point. The American version of the social justice claim is based on the federal government financially compensating the descendants of Black slaves. The African version is much more radical, demanding that all the nations that were net beneficiaries of the slave trade should compensate the African regions from which the slaves were taken, including the lands confiscated for purposes of imperial enrichment. Twenty years ago, when the US Gross Domestic Product was $9.7 trillion, the estimated cost of such reparations worldwide was $777 trillion (BBC News 1999). This enormous figure assumes that had the people and resources remained in Africa under native control, the amount of wealth produced would have been comparable to the wealth produced under colonial rule. Needless to say, this is a controversial assumption, which helps to explain why little action has been taken on the associated reparations claim.

My point is that general agreement on the nature and magnitude of a past injustice is insufficient to achieve policy closure on how to go forward. The justness of any proposed compensation scheme ultimately turns on counterfactual speculations about how the expropriated people and resources would have been disposed in a non-oppressed state. For many this is an uncomfortable and perhaps even 'politically incorrect' style of thinking, though it characterises the 'law and economics' movement in contemporary jurisprudence (e.g. Calabresi and Melamed 1972). Moreover, once social justice warriors are called upon to define the exact loss that has been suffered, they themselves are thrown into the post-truth condition. They are forced to relate the factual world in which injustices have occurred to an imaginary just world in which the harms had never taken place. Under normal circumstances, this would be the stuff of time-travel science fiction. But in practice it means that the social justice game must be played on the Rawlsian field of hypotheticals.

More generally, once you decide to fight injustice on 'truth' grounds, you are vulnerable to a pincer attack from, so to speak, *the whole truth* and *nothing but the truth*, which taken together could undermine the grounds on which you stake your own truth claims. On the one hand, your opponent may accept your facts but then introduce additional facts that alter if not undercut the normative force of your own facts. For example, one might enumerate various benefits of the slave trade, perhaps involving the complicity of the slaves themselves, which mitigate against the charges brought against the slave traders. On the other hand, your opponent may simply deconstruct the narrative that ties your facts together, implying that your account of the truth is designed to bias judgement in favour of your desired outcome. For example, doubts may be raised about whether the slave peoples were somehow 'held back' from a better future that they might otherwise have had, given the conditions in their homelands at the time of the slave trade. To be sure, such moves and countermoves are controversial, but they have become commonplace in humanistic scholarship about other matters over the past half-century.

However, the law does not enjoy the luxury of letting such disputes rage indefinitely. It must reach decisions over matters on which academics may never come to agreement. The relevant principles and procedures contributing to the law's sense of 'gamesmanship' include habeas corpus (i.e. the need to link a specific crime to specific suspects and specific victims), rules of evidence and statute of limitations. The last is especially important, since over time the perceived nature and severity of an original injustice may change in light of later events, which again may make it difficult to establish relevant facts or even a sense of fairness for purposes of adjudication. Indeed, the fundamental role played by statute of limitations in most legal systems testifies to a post-truth sensibility that concedes the inherently unstable factual basis of any claims of injustice. All of this is bound to be cold comfort for restorationists convinced that the 'Truth shall see you free' means that the wronged will be vindicated simply by allowing the facts to speak for themselves. Unfortunately, the facts are no less equivocal than the people who produce them, especially once the facts are emancipated from their original contexts.

Moreover, the prospects for the restorative justice agenda have been 'always already' undermined from within. The consolidation of multiculturalism around identity politics in the 1980s occurred against the backdrop of postmodernism, which had adopted a sceptical position towards the search for origins in general – of oneself, one's group or, for that matter, one's putative oppressors. Postmodernism 2.0 is alive and well in the current wave of 'trans-' sensibilities, which highlight the degree of discretion that people have to define themselves, depending on which strands of their complex past they wish – and are allowed – to project into the future. For example, shall they stick to their birth sex or express a 'biologically suppressed' sexual identity? This movement can be seen as an unintended consequence of the legitimisation of 'queer' identities in the academy where such people had already long flourished, albeit 'in the closet' (Sedgwick 1990). Thus, the accompanying sense of 'performativity' was initially meant in a descriptive sense – the sheer phenomenon of 'passing' – to highlight the socially constructed character of gender (Butler 1990).

But once they were out of the closet, previously hidden 'queer' practices acquired a cachet that made them attractive, more explicitly elective – and, most controversially, extendable into other dimensions of the human condition, including race and ethnicity. A high watermark has been the case of Rachel Dolezal, who was born White but identified so completely with Black oppression that she succeeded not only in passing as Black but also in becoming a leading Black activist. So had Dolezal simply become Black, full stop? Clearly the answer is not as clear-cut as one might have wished since the publication of a reasoned defence of Dolezal's racial 'transitioning' caused so many problems for a leading US feminist philosophy journal that several editors were compelled to resign (Tuvel 2017). One looks forward to the day when we come to see this episode as simply a moment of struggle that will accept free transit across identities as the culmination of the great *Liberal Revolution* that began in eighteenth-century Europe when the law took the first decisive steps towards establishing a 'level playing field' by restricting the scope of hereditary entitlements for those who enjoyed them and releasing those burdened by their heredity to become 'self-made' (Rothblatt 1995). The post-truth condition is usefully seen as the ultimate outworking of this world-historic emancipatory movement (Fuller 2019b: ch. 2).

12

A PANDEMIC SEEN THROUGH
A POST-TRUTH LENS

One can think about the COVID-19 global pandemic (active as this is written) in four orders:

1. *First order:* Winning the fight over the virus. This is defined nation by nation, and mainly for domestic consumption. Here one should expect considerable variation.
2. *Second order:* Winning the fight over what 'winning the fight' means. This is defined at an international level, probably by the World Health Organization. It may end up invalidating some of the first-order claims.
3. *Third order:* Winning the fight over the lessons to learn from winning the fight. This is defined by whether people want to go back to 'business as usual' with minimal disruption or take the crisis as an opportunity for 'no more business as usual'.
4. *Fourth order:* Winning the fight over what the lessons learned mean more broadly. This is how the crisis comes to define who we are – and in that sense, we come to 'own' the crisis as having made us stronger by not killing us.

These four strands of thought are 'ordered' logically, in the sense that the later ones presuppose some sort of closure on the earlier ones. Of course, in practice, all four orders are discussed simultaneously, though over time the different orders are gradually disaggregated to become associated with discussions about the 'past', 'present' and 'future' of the pandemic. What Hegel and others have called the 'logic of history' begins from this awareness (White 1973). Thus, 'first order' talk is most easily pinned down to events, whereas 'fourth order' talk is about what emerges as having been the pandemic's legacy.

When the COVID-19 crisis was about to send my university into lockdown in mid-March 2020, I sent a message to all my students about how they might think about the crisis (Fuller 2020). I stressed the anchoring effect of how the crisis first received worldwide attention. An overstretched doctor working in an intensive care unit at a hospital in Wuhan, China, panicked at the rate of patient intake and their rate of death – and did so on social media. The social media part was the truly new development, arguably more novel than the corona virus itself. After all, viruses mutate all the time and can even pass between species, and air pollution in Wuhan – like other big Chinese cities – had long been a breeding ground for the viruses that cause respiratory ailments. Indeed, it is an ecological legacy that has been felt in the waves of more or less lethal forms of 'influenza' worldwide that have beset the entire industrial era.

In the first instance, influenza tests the capacity of health care systems, and in the past most have struggled and many have failed to cope – but without raising the alarm at the global level to such an extent that capitalism was brought to its knees, as it has been by COVID-19. After all, it is not every day that the *Financial Times* (2020) calls for a 'Welfare State 2.0' on a scale that seventy-five years ago led to the establishment of the United Nations. Because flu viruses typically feed on existing vulnerabilities in a patient's condition, hospitals would typically need to register a higher than expected morbidity rate before a new virus was detected. But of course, previously the world's health care systems were not being exposed simultaneously for all to see on social media. As the capacity for a virus to go 'viral' increases, sensitivity to the virus's presence increases, which in turn reduces the time before an 'epidemic' and then a 'pandemic' is declared, which serves to accelerate the comprehensiveness with which measures are taken to fight the virus. Moreover, and importantly, intensified scientific scrutiny on the virus also generates back propagation effects, effectively a rewriting of medical history whereby earlier data is reclassified so that a greater number of people – both living and deceased – are revealed to have contracted the virus.

What I have just described is the outworking of a *quantum epistemology* whereby ailments and deaths that would have been diagnosed as extreme instances of the default disease categories are 'converted' – as in a Gestalt shift – so as to be directly attributed, both in retrospect and in prospect, to COVID-19 through the 'observer effect' of the intensified medical gaze (Fuller 2018: ch. 6). In short, the more we look for COVID-19, the more likely we are to find it. What had been non-existent is now and always everywhere. A similar observer effect will occur once the pandemic's conclusion is formally declared by the World Health Organization, and we properly enter the 'lessons to be learned' phase. There will undoubtedly be a reversal of fate for at least some policies that are currently being hailed or condemned, once the final toll has been taken. (I write this in early April 2020.) Indeed, the run-up to this second moment is marked by the sort of competitive expectation management that we normally see when 'bull' and 'bear' analysts try to forecast the stock market. Each wants to be the first to declare how things will have always been.

For example, based on the model used to guide the UK's COVID-19 strategy, the nation's Chief Scientific Advisor Patrick Vallance told a Parliamentary Select Committee at the start of the pandemic that 20,000 British deaths would be a 'good outcome'. Others were already writing post-mortems on the UK pandemic strategy before all the bodies have been counted. Critics have often made life easy for themselves by denying at once the validity of the target and the government's strategy to reach it – typically by pointing to the example of Germany, which was expected to have many fewer deaths. So even if Vallance's 'good outcome' had been met (within six weeks of his statement, UK deaths surpassed 20,000), the critics could have still accused the UK government of having 'cynically' set its expectations low. An interesting epistemological difference between the government and its critics is that the government stresses that the success of a particular nation's COVID-19 strategy will have been due to a combination of factors specific to that nation, whereas the critics stress that, say, Germany already possesses the all-purpose 'magic bullet' – mass testing of the population – that nations everywhere should have deployed to conquer the virus.

We might think of this difference as a matter of 'anticipatory diagnosis'. At a more general level, the market in expectation management is emblematic of the post-truth condition. At stake ultimately is not which nations tackle the virus most effectively but the terms on which 'effectiveness' is decided. If the virus is the opponent in a match against humanity, the main question is not who wins but who gets to name the game being played. It is a 'second order', not 'first order', problem, as the logicians say. This explains why the UK can't simply get away with setting its own targets, even if it meets them. Moreover, one can envisage 'third order' terms in which to deal with COVID-19. So let's imagine that the current fight against the virus is 'completed', at least in the sense that the spread of COVID-19 is 'contained'. This opens a market in which some are 'projecting' (i.e. both assuming and promoting) that 'we' (i.e. most of us) have won and others that we have lost. The former want to resume business as usual as quickly as possible, while the latter want to take the radically changed circumstances as an opportunity to play a different game. That is the third order game against the virus, on the cusp of which sits the *Financial Times* editorial mentioned earlier.

The editorial cites the United Nations (UN) as a precedent for a post-pandemic world order, but its history displays the ambiguous field of play in the third order game. Much of the organization's early rhetoric portrayed it as a proper world government that would establish human solidarity on a new footing, one that transcended the modern system of nation-states, known in international relations circles as the 'Westphalian Order'. Thus, the UN Universal Declaration of Human Rights alluded to the unprecedented horrors – 'crimes against humanity' – unleashed by the Second World War as the pretext for the UN Charter. Indeed, the very idea of 'human rights' with juridical standing pointed to the insufficiency of 'civil rights' inscribed in national constitutions. No more business as usual! Unfortunately, the Charter's details reveal that the UN's powers are no more than what its constituent nation-states are willing to delegate to it. The UN has no independent authority to be a proper rival to the nation-states themselves, and over the years it has best lived up to its early rhetorical promise when dealing with failed states and failed interstate relations. In effect, the UN is little more than the Westphalian Order's insurer of last resort: It has maintained business as usual.

What I have recounted may sound like a sorry tale. However, it is worth recalling that 'business as usual' is also the spirit in which national governments have bailed out banks following global financial crashes from the Great Depression onwards. To be sure, the bailouts have been cast in the confident rhetoric of building 'robustness' and 'resilience' into the system – but we're still talking about the same system that generated the crises in the first place. Nevertheless, those desiring to turn the latest global crisis into an opportunity for radical change – from Marxists to environmentalists – have been historically hamstrung by such rhetoric. In the 1930s radical socialists were dealt a sharp and arguably irreversible blow by the speed with which Franklin Roosevelt proposed and enacted the 'New Deal' legislation whereby the emerging superpower showed the world the way back to business as usual in the capitalist world order. It is not unreasonable to look to the current emerging superpower, China, for economic guidance in a post-pandemic world – especially since, like China today, the US had practised a notoriously protectionist international trade policy in the years prior to the Great Depression. But will it be 'business as usual', whatever that would mean?

Of course, it need not be. The alternative version of the third order game has not gone away. Its 'no more business as usual' attitude is predicated on our having lost the fight against the virus. However, this may not be as bad as it first seems because we're imagining this prospect, rather than having experienced it. That cognitive difference can effectively immunise us from the worst effects of the virus – or, indeed, any existential threat. Thus, China could provide a 'no more business as usual' style of global leadership post-pandemic by taking advantage of its authoritarian political structure and the depressed fossil fuels market to expedite a massive five- or ten-year plan to decarbonise its economy, a task that is generally seen as a long-term worldwide necessity. At once China could leapfrog ahead of the world in seeding an ecologically sustainable economic infrastructure while repairing the reputational damage caused by its having 'caused' the pandemic.

The modus operandi I have just described corresponds to how the great US Cold War strategist Herman Kahn (1962) thought about 'the unthinkable', namely, a thermonuclear war. Kahn estimated that the realistically worst outcome of a nuclear confrontation between the US and USSR would kill no more than a third of the human population. And without denying such a tragic loss of life at an unprecedented scale, his strategic focus was on how the survivors might coordinate their activities to rebuild civilisation. After all, the existing communication infrastructure would be among the most vulnerable targets of such an all-out war. This was the context in which the internet was developed by the US Defence Advanced Research Projects Agency (DARPA) in the late 1960s. Of course, the imagined war never happened, but the internet was rolled out anyway – and became the platform on which the 'third industrial revolution' was launched (Rifkin 2011).

I have characterised this sort of third order response as 'moral entrepreneurship', in the spirit of Obama Chief of Staff Rahm Emanuel's notorious offhand comment, 'Never let a good crisis go to waste' (Fuller 2012: ch.

4). More generally, its attitude to risk is not precautionary but *proactionary* – which is to say, it capitalises on the nature of uncertainty to convert potential threats into opportunities (Fuller and Lipinska 2014). As Franklin Roosevelt put it in his 1933 inaugural presidential address, 'there is nothing to fear but fear itself'. However, the sensibility goes back to Baron Helmuth von Moltke's successful revision of strategic thinking during the 1870–1 Franco-Prussian War. Moltke effectively raised the stakes of warfare from limited objectives to national security itself, always with an eye to the next war – and there will be a next war, perhaps against some radically different opponent, even if one wins the current war against the known opponent. Moltke's vision was one of endless 'total war', which required the nation to be in a state of 'permanent emergency' to ensure that, whatever the outcome of any particular war, the nation is never defeated 'once and for all' but can always 'rise again' (Fuller 2000b: 105).

Kahn and other Cold War strategists adopted Moltke with gusto, resulting in a raft of new sciences and technologies – as in the Franco-Prussian War, when epidemiology came into its own as a science, with Louis Pasteur and Robert Koch spearheading their respective national biomedical efforts (Fuller 2018: ch. 4). Philip Tetlock's concept of *superforecasting*, which duly acknowledges Moltke, is the latter-day incarnation of this mentality (Tetlock and Gardner 2015: ch. 10; Fuller 2018: ch. 7). But perhaps most consequential of all is what might be seen as a 'fourth order' effect of the original fight against the virus. It came to be associated with Moltke's civilian commander, Otto von Bismarck, who turned the Second German Reich into the first modern welfare state – what the US sociologist Alvin Gouldner (1970) pointedly called, at the height of the Cold War eighty years later, the 'welfare-warfare state'. Bismarck fully conceptualised the nation-state as an organism perpetually struggling for survival in a potentially hostile environment. Protection literally meant education and health coverage to keep everyone 'fighting fit' to contribute to economic productivity in peacetime and military might in wartime. Its modus operandi included taxation and conscription.

But Bismarck did more than that. His specific 'fourth order' genius lay in analogically extending the idea of 'potential external foes' to include radical new ideologies, such as Marxism, which had already threatened to undermine the nascent welfare state by portraying it as an engine for generating an endlessly exploitable population. Bismarck's solution was to introduce retirement pensions, showing that people were valued even when they were not directly contributing to national security. This resulted in a comprehensive 'social security' system that extended beyond the borders of the nation-state to its constituent human members. In terms of the *Realpolitik* of which Bismarck was the master, this policy neutralised many Marxists who ended up normalising themselves as the Social Democratic Party, which became the largest party in the Reichstag and even supportive of Germany's ill-fated involvement in the First World War (Bew 2016: ch. 3). Political sociologists recall this episode as the paradigm case of *co-option*, the means by which the body politic establishes equilibrium with a potentially lethal enemy by incorporating it as a kind of 'natural vaccination' (Michels 1959). In effect, Germany achieved 'herd immunity' with regard to radical left-wing politics, notwithstanding Marx's own prediction that due to its large organised labour movement the proletarian revolution should first happen there.

Arguably the last crisis to have substantial fourth order effects on human self-understanding was the 1755 Lisbon Earthquake. Based on the records of the time, seismologists estimate that it registered 8–9 on the Richter scale, which is unprecedented in living memory. (The California earthquakes of the past fifty years have registered 6–7 and even the 1995 Kobe, Japan earthquake registered only 7.) Moreover, Lisbon was still a major city in the mid-eighteenth century. Nevertheless it lost at least a quarter of its population and Portugal nearly halved its GDP. The earthquake's shock waves were felt as far away as Finland, North Africa and possibly the Caribbean. The major philosophers of the time – including Voltaire, Rousseau and Kant – wrote extensively about it. Generally speaking, it shook up not only faith in God's benevolence – if not his very existence – but also faith in humanity's own ability to tame the forces of nature through 'civilisation'. These doubts about God and humans capture Voltaire's and Rousseau's respective contributions. Voltaire left the greater literary legacy, the satirical *Candide*, while Rousseau cast the longer shadow on the entire Enlightenment mindset that they originally shared. As for Kant, his essays on the earthquake introduced the metaphor of 'firm/shaky foundations' (*Grund*) for knowledge claims, which have become a staple of philosophical and even popular discourse in the modern period. Kant's contribution proved to be the ultimate fourth order effect of the crisis.

13

THINKING IN THE FOURTH ORDER: THE ROLE OF METALEPSIS IN THE POST-TRUTH CONDITION

For the current pandemic to have a similar fourth order impact, it would need to be lifted out of the realm of metaphor. More precisely, the pandemic would need to become *metaleptic*. Metalepsis is the rhetorical trope most closely associated with 'sublimation', the term that Freud artfully repurposed from chemistry, where it referred to the passage of a substance from a solid to a gas state without having passed through a liquid state. For Sigmund Freud (1930), the trappings of civilisation – including art and science – are the collective projection of the individual superego's internalisation of the parental voice. In other words, a distant personal memory that is presumably shared by everyone becomes the standard by which their collective endeavours are pursued and judged, without that standard ever having been subject to proper argument or debate. Put in most general terms, metalepsis flips the metaphorical and the literal. What started out as a purely imaginative attempt to make sense of reality comes to be interpreted as its deep structure. This sort of radical transformation is especially evident in the history of science, which is embarrassingly full of counter-intuitive – if not outright preposterous – ideas that end up redefining our understanding of reality.

Perhaps the easiest way to make this point – though interestingly not a route pursued by most historians – is to chart the travails undergone by the *atom*, whose existence was agreed by scientists only in the early twentieth century, notwithstanding its long intellectual history. Afterwards the ancient atomists Democritus and Lucretius were increasingly described not simply as philosopher and poet but as 'proto-scientists', even though neither thought an experiment was necessary to demonstrate their claims. Meanwhile the competing entity *energy* retreated from the fundamental ontology of natural science to the interdisciplinary purgatory known as 'ecology' or 'environmental science', where it still has purchase among its hybrid natural-social scientific inhabitants. The legacy for us living in the early twenty-first century is that atoms are definitely real – and energy, well, less so. Moreover, we treat this judgement as applying not only to how we go forward in the future but also to how the past had been. Thus, whatever their other virtues, those past scientists who defended energy over atoms are deemed on the wrong side of history, and often pay a heavy price. For example, Wilhelm Ostwald, a Nobel Prize-winning founder of modern physical chemistry who converted his belief in energy into a briefly popular world view ('energeticism'), is now judged a historical curiosity. A great German phrase – *wird nie passiert sein* ('will never have happened') – captures the totalising 'Gestalt' shift in horizon described in the fates of energy and Ostwald, which is tantamount to being written out of history.

To understand the work of metalepsis, it is worth recalling that metaphors are usually intended to be quite limited in scope, which is what makes them appear striking yet also 'merely' figurative. 'The early bird catches the worm' works as a piece of wisdom just as long as one doesn't dwell for too long on how the connection between, on the one hand, birds and worms and, on the other, humans and their goals is supposed to work. The 'magic' of the aphorism lies in its constructive ambiguity: the listener has considerable discretion in how s/he completes the meaning of what has been said. To be sure, some – especially evolutionary psychologists – may believe that the underlying causal process in both the avian and human cases is substantially the same. In that case, the metaphor acquires the status of an analogy, which is then pursued scientifically in terms of more detailed possible correspondences between avian and human behaviour, genetics and so forth.

The fate of that exercise is normally an empirical matter, with the expectation that the analogy will prove true in some respects but not in others. However, sometimes in the course of inquiry the two sides of the analogy effectively reverse roles. Thus, instead of asking whether birds can explain what humans do, whereby humans set the standard that needs to be met, we simply presume that birds explain what humans do, in which case the burden of proof is placed on the humans to explain why they deviate from avian expectations when they do. At that point, the metaphorical is rendered literal, and we enter the intellectual wormhole that is metalepsis.

Deep conceptual revolutions are metaleptic. The world literally comes to be seen from an entirely different point of view. We know that it is 'an entirely different point of view' because something that had previously been an object of our knowledge is now constitutive of how we know the world. We don't see the sun as something separate from us; rather we see as the sun. The 'other' has come to be incorporated as part of our lifeworld. Thomas Kuhn (1970) famously caught a glimpse of this insight when he likened revolutions in science to Gestalt shifts. The same data come to be seen as organised in a radically different way, in which the change of perspective amounts to a change in world view. Kuhn's main historical example was the Copernican Revolution, which involved learning to make sense of the heavenly motions as if one were not planted on Earth but rather observing from the sun. Newton then extended this horizon from the here and now to anywhere and everywhere in space-time, thereby making explicit the long simmering idea that humans could see themselves not simply as animals planted on Earth but as beings capable of adopting God's point of view (Koyré 1959).

The Harvard historian of science Gerald Holton (1973) rendered Kuhn's insight more concrete by reviving a late nineteenth-century German conception of scientific theories as *Weltbilder*, or 'world constructions'. If we believe with Newton that the 'world is a machine', albeit one created by God, then 'machine' is not simply a metaphor or even an analogy that generates testable hypotheses. It is more than that: it is a normative standard to which we hold the world accountable. That which does not behave mechanically becomes deviant and potentially problematic because it violates the terms of what is possible – or more to the point, permissible – in this *Weltbild*. This is the sense in which the so-called Newtonian world view was dominant from the early eighteenth to the early twentieth centuries.

What Kuhn called a 'scientific revolution' occurs through a systematic reversal of metalepsis. Once the dominant 'paradigm' (Kuhn's proxy for *Weltbild*) has accumulated enough unsolved problems over a long period, it precipitates a 'crisis' whereby what had previously functioned as a norm is rolled back to an analogy whose hypothetical status then renders it vulnerable to still further reductions in its epistemic status. After a while, the field is opened to replace the paradigm, one of which metaleptically emerges as the 'new normal' for the world that will have been constructed. In effect, we wake up from one dream, only to be captured by another. And so Newton's mechanism came to replace Aristotle's organicism, only to be itself replaced by Einstein's relativism. Arthur Koestler's (1959) *The Sleepwalkers*, which depicted the rise of modern science as an eruption of the unconscious, may have been right, after all.

The idea of *Weltbild* is itself inherently ambiguous with regard to the relevant sense 'construction' – or 'building', etymologically speaking. When this term first circulated among physicists, it was in the spirit of a 'world picture', a systematic image of the world. It was associated with *Weltanschauung* (world view), which is the modern idea of 'aesthetic perception' that Kant adapted for his own purposes in the *Critique of Judgement* shortly after its coinage in the late eighteenth century. However, the unprecedented carnage of the First World War resulted in much literal 'building' of European infrastructure and institutions – often from the ground up. In this context, Rudolf Carnap and his fellow logical positivists began to interpret *Weltbild* more concretely as *Aufbau*, a term that Niels Bohr had recently brought into scientific usage to describe how matter was built up from atoms, but which also had currency in the ongoing modernist (*Bauhaus*) revolution in architecture (Galison 1990). The positivists spun the meaning of *Aufbau* to comport with Kant's earlier metaphorical innovation of 'foundations' from the Lisbon earthquake.

Twentieth-century philosophy's signature preoccupation with establishing 'firm foundations' for knowledge, morals and so on, is a consequence of this particular feat of metalepsis. Earlier philosophers – even ones like Descartes who sought certainty in all things – did not envisage the task in such concretely 'foundationalist' terms. Nevertheless, in the spirit of a Gestalt shift, Descartes and most of the same historical figures continued to be discussed after this revolution in philosophy – except that their respective strengths and weaknesses as thinkers looked somewhat different, sometimes radically so. What is perhaps most striking about 'revolutions' in this sense is the relatively little effort needed to produce a relevantly large effect. To recall the original Gestalt psychology experiments, subjects delivered radically alternative judgements of what they were seeing, depending on the contextual cues given in the face of an ambiguous object. Indeed, to change one's mind is arguably no more than to rearrange the relative significance of what one already knows. In that respect, 'evidence' is just a euphemism for the trigger moment when it happens.

If the crisis surrounding COVID-19 holds the same metaleptic potential, what might that be? To be sure, the idea that a fatal illness might have an alien airborne source is far from new. The etymological kinship of 'influenza' and 'influence' in the early modern period provides a natural start to this story. However, originally the discussion was not metaphorical at all. People actually thought that the motion of the heavenly bodies (somehow) caused people to become ill, depending on their birthdays. Moreover, this 'illness' was conceived in psychosomatic terms – that is, not confined to either the mind or the body. Such were the ways of astrology. As astrology faded as an acceptable *Weltanschauung*, 'influenza' and 'influence' came to be associated with different causal streams, effectively reduced to 'mere' metaphor. Psychoanalysis has been arguably the strongest 'scientific' standard bearer of the old astrological line over the past five hundred years. (Here Carl Jung deserves 'full marks'.) Otherwise the default tendency has been for a greater disaggregation of 'influence' and 'influenza', periodically punctuated by controversy whenever someone seemed to allege a closer connection, as when Richard Dawkins modelled 'meme' on 'gene' in *The Selfish Gene* (1976). However, a close reading of Dawkins reveals that the real model for the meme was the *virus* – that is, free-floating strands of DNA that require a host to replicate and potentially alter the host's genetic make-up. And so the move to metalepsis might begin.

Here it is worth recalling that the 'virus' as an entity distinct from 'genes' or 'germs' dates no earlier than 1898 (due to the Dutch botanist, Martinus Beijerinck), despite various premonitions that such an entity might exist. And even that was a half-century before the DNA revolution in molecular biology could begin to explain viruses properly. But perhaps more to the point, it was long before viruses began to be engineered by biomedical scientists to produce potentially permanent changes in the genetic make-up of organisms. The spirit of the activity is one of simulating what is known in evolutionary theory as 'horizontal gene transfer' whereby genes are transmitted between two species that are not directly related in conventional taxonomic terms – such as bat-to-human, in the purportedly original case of COVID-19 in Wuhan, China. Of course, there is an obvious difference between what genetic engineers in laboratories do and what nature does when mutating viruses spontaneously. The former is deliberate and therapeutic in intent, while the latter is arbitrary and indifferent – at least from a human standpoint. Nevertheless, it would be difficult to draw a clearer line of battle – marked by the virus – between humanity wanting to turn nature for its own purposes and nature having a mind of its own.

Insofar as genetic engineering will increasingly feature in humanity's medical armament, nature can and will use the same weapon against us. Those are the battle lines of any future virus wars. In the late nineteenth century, Louis Pasteur famously put the entire French nation on a war footing with regard to 'microbes' as the 'invisible enemy'. His metaphor stuck and was extended across biomedicine in the twentieth century in which many diseases became 'silent killers'. However, the occurrence of a pandemic at this stage in scientific history potentially carries metaleptic import because we now know enough about viruses to be able to generate them not simply by accident, but on purpose. Little surprise, then, that US President Donald Trump among others have floated the idea that COVID-19 was somehow 'manufactured' in Wuhan. In any case, it put long-standing ecological concerns about humanity's relationship to nature on a much more literal footing, rendering it 'up close and personal'.

In conclusion, let me underscore why the intellectual trajectory I have been outlining to navigate through the COVID-19 crisis is 'proactionary' rather than 'precautionary'. Adherents to the precautionary principle tend to presume that humanity is 'always already' subordinated by some generic external agent called, say, 'Nature' or 'Gaia', in terms of which we might be either in equilibrium or in conflict. In either case, that opponent is not subject to negotiation: it poses an ultimate limit of permissible activity. In contrast, adherents to the proactionary principle tend to presume a more gamelike relationship to the opponent, whereby Nietzsche's Zarathustrian imperative, 'What doesn't kill me, makes me stronger', really does apply. This suits the world of endless pandemics in which the opponent may be contained for now but may mutate to pose a future threat, which in turn may require a radical revision of one's modus operandi and even raison d'être – including an incorporation of the 'other' previously not seen. After all, we should never let a good existential threat go to waste.

14

THE PATH FROM FRANCIS BACON: A GENEALOGY OF THE POST-TRUTH CONDITION

Lying is the spontaneously human activity on which the post-truth condition is based. It requires a mental division between what in the modern period has been identified with the 'public' and 'private' expression of thought, in which either can be judged in terms of the other. But in both cases, a distinction is implied between 'first order' and 'second order' thinking whereby the former is judged by the latter. But which is which in practice? Consider *hypocrisy*, a term adapted in the early modern period from the wearing of masks in Greek drama. A 'hypocrite' is no more than someone who says what needs to be said publicly, regardless of its conformity to one's personal beliefs. While this looks like an open invitation to lie, it may also compel the person to tell the truth in a way that one could not have done, had one spoken in a 'private' voice (Fuller 2009: ch. 4). This ambiguity fascinated the founder of the scientific method, Francis Bacon, who, as personal lawyer to England's King James I, crafted an 'art of experiment' to test potentially all knowledge claims, based on the 'inquisitorial' mode of judicial procedure in continental Europe (Fuller 2017).

Bacon gave lying its due. He realised that witnesses lied regularly, yet the conditions under which they feel compelled to lie may still reveal hidden truths – in which case their lies may not to be so bad, after all. They may even be productive. In ordinary court cases, all depends on the exact game that the inquisitor is playing with the witness in a given trial. Nevertheless, Bacon concluded that the sort of elaborate artifice – both human and material – that nowadays routinely accompanies the conduct of scientific research is best suited to uncovering nature's secrets and perhaps even deciphering the signature of their divine author. Here, Bacon benefitted from facing two centuries at once. To be sure, he is normally seen as facing the seventeenth century, in which he appears as a fellow-traveller of Galileo in pioneering the scientific method. But he equally faced the sixteenth century, in which he figured as a Renaissance essayist who rivalled his older contemporary Michel de Montaigne's capacity to seamlessly interweave sacred and pagan sources, which provided the stylistic basis for modern prose. It is through Bacon's encounter with Montaigne that Adam's lying came to pave the way to secular modernity.

Bacon's famous essay, 'Of Truth', alludes to Montaigne's brief discussion of Adam's fall in the latter's 'Of Giving the Lie'. One point on which they agreed was that Adam offended God less by eating the forbidden fruit than by denying the deed after the fact: that is, by lying. It is this interpretation of Adam's transgression that had led Augustine to formulate the doctrine of Original Sin early in the history of Christianity. The part of Augustine's doctrine that people remember is that every subsequent human generation is tainted with Adam's transgression. It amounts to a permanent debt that humanity must carry until further notice, as reflected in the drudgery and mortality of our everyday lives. However, it is often forgotten that our free will – the feature that makes us most like God and least like animals – remains intact even after Original Sin. In effect, God continues to allow us to transgress if we so choose: we retain the right to be wrong and the freedom to make our own mistakes – and to lie.

For a long time, Original Sin was regarded by the Roman Catholic Church as a curious and rather extreme doctrine. It seemed to exaggerate both the heights from which Adam had fallen and the depths to which he had sunk. Yet by the early modern period, under the influence of Protestantism, Original Sin had become one of the main grounds on which Christianity was distinguished from Judaism and Islam – perhaps second only to the divine personality of Jesus, and in fact related to it. These other Abrahamic religions accept that Adam disobeyed God but do not accord any special moral significance to his prevarication about it. Because Jews and Muslims do not recognise the divinity of Jesus, they are not compelled to commit to the idea that humanity partakes of specifically divine qualities such as absolute truthfulness, even if God privileges us above all the other animals.

The crucial point here is that Judaism and Islam do not confer on human language the sort of godlike creativity that could make lying metaphysically dangerous. On the contrary, Jews and Muslims regard God's relative indifference to human lies as indicative of the deity's supreme magnanimity in the face of inherent human weakness. After all, our lies do not prevent God from knowing what we seek to conceal. To be sure, such a relaxed attitude to lying has played into modern 'orientalising' stereotypes of Judaism and Islam as somehow 'loose' or 'decadent' because their deity would seem in the end to forgive virtually anything that humans might say or do. So, what exactly is the Christian problem with lying – and what is its legacy for our secular times?

It is interesting to think about this question in light of Bacon and Montaigne, neither of whom can be regarded as conventional Christians. Bacon developed the scientific method out of his sympathy for the 'magicians' whose practices had been banned by most Christian churches, while Montaigne's preoccupation with humanity's various animal-based weaknesses have led many readers to wonder whether he really believed in an immortal soul. Nevertheless, both clearly resonated to Augustine's doctrine of Original Sin. They were not drawn to the popular Catholic idea that Adam lied to God out of shame for his transgression, which implies a sense of recognition and perhaps even remorse for his error. (This is the figure of Adam holding a fig leaf over his private parts.) For their part, Montaigne regarded Adam's lying as demonstrating 'contempt' for God, while Bacon more euphemistically described it as 'brave'. It would seem that God was compelled to humble Adam because Adam refused to humble himself. Adam and Eve's expulsion from Eden was, therefore, the outcome of a battle of wills.

This general sense of defiance would soon be found in Milton's portrayal of Satan in *Paradise Lost*, which in turn contributed to the revival of the Greek legend of Prometheus, himself the product of divine and human heritage who steals fire from the gods to give to humans. (Here 'fire' stands for a general principle of change, the capacity to turn one thing into something else.) In the Romantic period, Goethe's *Faust* and Mary Shelley's *Frankenstein* popularised this image of humans as beings who would arrogate to themselves a sort of knowledge that is normally only God's, albeit with little understanding of all the relevant consequences. A subtle yet enduring legacy is the inversion of the meaning of 'innovation' in the nineteenth century. At the start, it referred to the monstrous corruption of ancient wisdom, but by the end it had come to mean the marvellous creation of a new truth. The shift amounted to an admission of humanity's godlike capacity for original creativity. The inventions that were the basis for these innovations – typically machines – came to be seen not as better or worse forgeries of nature but as creatures in their own right that are entitled to their own form of protection, to which we nowadays often attach the phrase 'intellectual property'.

Writing at the end of the nineteenth century, Nietzsche could easily see in this line of thought what he called a 'transvaluation of all values'. But such a transvaluation had already been presaged in Montaigne's famous saying that the true is one but the false are many. The difference is that Nietzsche was placing a clear positive interpretation on Montaigne. Lying is effectively 'transvalued' from signifying the absence or deprivation of truth to being the generative source of alternative and even competing truths. There is a *logical* and a *genealogical* way to understand this transvaluation.

In logical terms, the nonidentity between the one truth and the many falsehoods is only partial: the multiple contradictions do not amount to a single contrary. In the end, Satan is not the anti-God. He is a delinquent creature of God. In the end, Adam defied God only on one point, but that nevertheless turned out to be one point too many. Because in most respects we may remain loyal to the truth, lies can easily pass as truth. In genealogical terms, the many falsehoods owe their existence to a progenitor truth from which they deviate. This insight lay behind Nietzsche's claim that modernity consists in humans transitioning from being without God to becoming godlike: we shall occupy the space of God, as the firstborn occupies the parental estate – uncomfortably yet necessarily. This is Nietzsche's theory of the *Übermensch* in a nutshell. And so, our lies become the new truth, and our artifices – the 'innovations' – become the new furniture of the world, replacing that of God's nature. Indeed, in this brave new world, God is put at a distinct disadvantage, which is revealed by the sort of public relations that is increasingly done on his behalf in the modern era. God is presented less the fecund source of all being than as the judge who finally stops the fecundity of the human liars and artificers who normally plague the world.

It is worth observing that classical pagan culture and those early moderns who drew on it for their inspiration – from Plato to Machiavelli – were never forced into this drastic faceoff between God and humanity. They approached lying differently. For them 'knowledge' and 'power' are correlative concepts concerned with control over the truth. Indeed, in this way of thinking, absolute knowledge and the monopoly of power are the two complementary faces of truth. However, if the dominant party needs to engage in excessive force or even excessive arguments, then its implied control over truth is potentially weakened, enabling a 'false' pretender to claim a kind of legitimacy vis-à-vis the truth. This helps to explain Plato's policy of pre-publication censorship instead of public criticism in his ideal republic, and why Machiavelli believed that the best prince keeps the peace by creating a climate of fear self-imposed by subjects who imagine the consequences of disobedience. In both cases, the goal is to maintain the true by preventing the false from ever surfacing. The strategy is to ensure that force rarely – ideally never – needs to be openly applied. In this respect, political competence operates in perpetual deterrence mode, displaying a calm but fierce exterior. In science, this passes for 'peer review'. Thus, Machiavelli likened the guardians of knowledge and power to lions. Yet in the end he shared Plato's fundamental pessimism about their long-term success. And interestingly, just like Plato, Machiavelli diagnosed the problem mainly in terms of the inherent corruptibility of those who would assume the lion's mantle. Even those on top are ultimately floored by the baseness of human nature.

This is strikingly different from the Augustinian framing of the situation, which Montaigne and Bacon shared. For them the problem is not – as it would seem to Plato, Machiavelli and perhaps Nietzsche's *Übermensch* – that God might turn out to be some classical leonine autocrat who fails to respond adequately to human defiance. Rather, the confrontation between God and humanity might unleash what is most godlike in humans, resulting in an endless proliferation of alternative truths and the associated confusion of judgement and action across the entirety of Creation. This is certainly the spectre conjured up by Milton's Satan, as well as the argument that Milton himself pursued in his landmark 1644 tract against pre-publication censorship, *Areopagitica*. What we now valorise as Milton's defence of free expression was envisaged even by its author as capable of licensing open intellectual warfare that could result in violence and even death, as everyone exercised their godlike capacity to create through the word. In Milton's 'free' world, one person's *logos* may well turn out to be another's lie. When people nowadays fear the worst of our 'post-truth condition', it is a secular version of this scenario that they have in mind. The fear is *not* that people can't tell the true from the false, but that they cannot agree on the standards by which to tell the difference.

This idea of lying as the wilful defiance of established truth has left an indelible mark on the character of modern art. Its most obvious and articulate presence may be Oscar Wilde's 1891 dialogue, 'The Decay of Lying', which argues that the aesthetic quality of a work should be judged by the extent to which its own sense of 'realism' deters audiences from asking whether the art measures up to some other 'real world' standard. If so, the false is effectively indistinguishable from the true, rendering art self-validating – or 'art for art's sake', as Wilde himself memorably put it.

Wilde's line of argument recalls that used by Christian natural theologians to establish at once the existence of God and our knowledge of God. It amounts to saying that nature works as well as it does because it has been designed to work that way, and that any further questions we might have – say, about why certain aspects of nature don't seem to work so well – should involve understanding the designer rather than doubting that the design is really there. Wilde's blasphemy, of course, is that he would allow the artist to occupy the position that the theologians had reserved for God alone.

To understand lying as a sort of 'alt-truth' process was scandalous in Bacon's day and remains so in our own. Nevertheless, the sixteenth-century reappraisal of Adam's defiance of God's authority sowed the sense of human empowerment that came to characterise modern art, science and politics. In an ironic twist to Plato, this development shows that knowledge and power are indeed correlative concepts, but we have so far really only come to terms with democratisation of power but not of knowledge. And on this latter point, lying may provide a useful guide.

CONCLUSION: HOW TO PUT YOURSELF IN THE POST-TRUTH FRAME OF MIND

The post-truth condition encourages a deeper blurring of the boundary between fact and fiction than what has been associated with the conventions of *literary naturalism* that have been dominant in the West for the past 150 years. I have in mind the following writing strategies, which involve the authors fabricating evidence to drive their plots within the bounds of empirical plausibility:

1. Filling in the otherwise undocumented details of the lifeworld of historical agents, including their moral psychology, in an attempt to understand why they behave as they do – as in the novels of Emile Zola.
2. Presenting a fictionalised 'ideal-typical' historical agent who is a composite of real agents but drawn in high relief to allow the agent's essential character to play itself out openly – as in the novels of Thomas Mann.
3. Placing fairly ordinary people in extraordinary circumstances, yet ones that correspond to situations in which science and technology might place people in the future – as in the novels of H. G. Wells.

Wolf Lepenies (1988) has astutely linked these conventions with the emergence of the 'sociological imagination' in the late nineteenth and early twentieth centuries, as expressed in a range of novel empirical methods from investigative journalism and on-site ethnographies to mass social experiments. From that standpoint, the aim of sociology may be seen as transforming the novelist's individual fabrications into a collective discovery procedure and validation process that results in scientific knowledge. This transubstantiation of fiction to fact passes through stages of idiosyncratic speculation, hypothesis formulation and testing in a contrived environment. At each stage, a larger and more heterogeneous collection of people and interests are gathered, culminating in a published finding: a 'fact' observable and usable by all.

Bruno Latour (1987) has made a successful career out of describing this process of scientific fact construction in terms of the extension of networks, but he has done so by pulling his punches. The dynamic is really that of a 'self-fulfilling prophecy' (cf. Festinger et al. 1956). Science, simply by virtue of its ongoing mass performance, becomes the fiction that is 'too big to fail', as just too many people, resources and sheer effort come to be invested in the 'common task'. This last phrase, which comes from the Russian Cosmist philosopher Nikolai Fedorov, was refashioned by Lenin for his own scientific updating of socialism (Bernstein 2019). It set the pace for the Soviet Union's unique single-mindedness in advancing 'humanity' as a collective project, something which fascinated H. G. Wells throughout his life (Gray 2011: ch. 2). Wells certainly took this proposition seriously when he pitched his candidacy for the founding chair in sociology at the London School of Economics. Wells proposed a 'science of the future', invoking such pre-academic 'sociologists' as Auguste Comte, Karl Marx and Herbert Spencer as his precursors (Lepenies 1988: ch. 5). Wells's failure to secure the appointment consigned those inspired by his example to dwell in the intellectual purgatory known as 'science fiction'. Sociology meanwhile became the discipline of mixed empirical methods and strangulated theoretical horizons that we know today. However, the post-truth condition permits a 'deeper blurring' than what even Wells had envisaged.

Central to the post-truth condition is the struggle over *modal power*, which is ultimately a game over what the players regard as possible. Plato's signature contribution was to realise that power is most effectively exercised by those capable of conveying what must be the case in the guise of knowledge. Thus, once told of the path they must follow, the people come to believe that were they not to follow that path, chaos would befall them – not by the hand of Plato's philosopher-king, but by the hand of some unaccountable but no less 'necessary' sense of 'reality'. The people's simple belief in the prospect of chaos turns out to be enough to deter disobedience. By wielding power in the form of knowledge, Plato's philosopher-king minimises the need to take violent action against his subjects, which may prove self-defeating in the long run. Persuaded of their own limited modal horizons, the populace does most of the work for the ruler. When Jean-Jacques Rousseau began the *Social Contract* by declaring, 'Man is born free but everywhere he is in chains', he meant exactly this. We are enslaved by our own reluctance to imagine that things could be other than they are.

But what if people are actively encouraged to imagine the *necessary* as *contingent*? In other words, a foregone conclusion starts to be seen as dependent on something hidden, which once revealed would reorient one's horizons, perhaps because it could be easily removed or even reversed. One wouldn't need to deny the events themselves but only the narrative logic configuring them, which is understood as having been spun with an eye to constraining future action. This is perhaps the best context for under-standing the post-truth proclivity to conspiracy theories, as these too turn on the prospect of strategically withheld information designed to give people a false impression of how some event happened. But a conspiracy need not be involved in the omission. The original observers to the events might not have been attentive in the right way to catch all the salient details. By rendering the necessary contingent in this fashion, our future horizons are altered because our sense of how we got to where we are has changed. We have effectively performed a Gestalt shift.

This process draws on what philosophers call a 'counterfactual' analysis of causation. The idea here is that very often we presume that because we believe that A was followed by B, B would not have happened were it not for A – and indeed only A. In other words, B could not have happened in any other way. But of course, B might have happened anyway, by some other means, with or without A. Similarly, it may be that B followed from A only because some other event C also took place – and C is what really matters. A by itself would not have been sufficient to bring about B. The playing field of modal power is levelled once these possibilities are taken seriously and exploited. By 'exploited' I mean that we really only ever know what we are told or can inspect for ourselves. Our belief that A 'caused' B is increased by our attending exclusively to the relationship between A and B. But of course, there is always more to find out 'behind the scenes', as it were, more details to fill in that may end up altering the entire look of a train of events, thereby questioning whether the route to B was governed by an ironclad narrative logic.

This mode of attack, once focused on the legality of the Catholic Church's secular entitlements, served to undermine established authority across the board in the European Renaissance. Were the original agreements between sacred and secular leaders as freely and transparently undertaken as we had been led to believe? Such a suspicious turn of mind corresponded to the rise of a more 'critical' and 'scientific' mentality that soon engulfed even the Church's spiritual entitlements in what became the Protestant Reformation. Founding documents were examined as evidence for the terms on which they were agreed – only to be frequently found wanting (Grafton 1990). This experience was the rite of passage to modernity, and it helps to explain the seventeenth- and eighteenth-century fixation on specifying the 'terms and conditions' on which any complex social arrangement is maintained, aka the 'social contract'.

As a literary experience, the necessary is rendered contingent by adopting a 'defamiliarised' standpoint to the default narrative logic. In practice, what is left unsaid in the narrative is treated not as commonplace, and hence justifiably omitted, but as pointing to something crucially missing, which if provided would change entirely how one sees the narrative unfolding (cf. Fuller 1988: ch. 6). Early in this book, I invoked a theological controversy that erupted at the turn of the last century that drove this point home for Christians. The controversy did not require denying any of the basic facts of Jesus's life. However, it did involve suspending an implicit assumption of the narrative connecting those facts that had been shared even by those who questioned the divinity of Jesus: namely, his sanity. Suddenly, the old and seemingly unresolvable conflict over Jesus's divinity was cast in a radically different light and, for a while, sidelined altogether.

It is worth emphasising how the above involved a deeper blurring of fact and fiction than, say, Zola's celebrated literary naturalism. The psychosocial details supplied in Zola's novels served to reinforce the overdetermined nature of his characters' existence. When sociologists today identify a family as living in 'poverty' and then proceed to show how that objective condition shapes their self-understanding and relationships, they are factually following in Zola's fictional footsteps. But the post-truth condition is about the opposite of this. In fact, it is closer to what Zola himself exemplified so brilliantly in his 'J'Accuse!' editorial in response to the conviction of Alfred Dreyfus for treason. Based on his own reading of the trial's transcript and the documents provided as evidence, Zola concluded that Dreyfus, as a Jewish army officer, was being conveniently scapegoated by the French Foreign Ministry for its own traducing of the national interest. Zola's suspicion was eventually proven well founded, but through no additional effort of his own. All he had to do to turn the tide was to systematically reinterpret the texts as concealing something significant, which served to sow the seeds of doubt in the case against Dreyfus. At the time, Zola's imaginative intervention crystallised the figure of the 'intellectual' in the public sphere (Fuller 2009: ch. 3).

I call Zola's intervention 'imaginative' because it was no less fictional than his novels. But it was a fiction that resulted in Zola's conviction for libel, which forced him into exile in London until Dreyfus's own conviction was overturned. At the same time, Zola's intervention was no *more* fictional than the cases mounted by the opposing lawyers in Dreyfus's trial. In any case, when Plato regarded 'poets' as the most subversive figures in his ideal republic, he had in mind just this ability to throw the rules of the game into doubt. Here it is worth recalling the etymological roots of poetry in *poiesis*, which is best translated as 'creativity', in the strong sense of bringing something into existence out of nothing – a filling of the void, a giving of voice to what is unvoiced. After St Augustine, *poiesis* constituted the divine signature for Christians, which with the spread of literacy was slowly welcomed as a feature of human writing. However, Plato was specifically threatened by the *poiesis* involved in the staging of scripted performances. They generate an alternative reality which, with the audience's participation, could become the dominant reality and thereby destabilise the social order. We call such performances 'drama'. Plato's misgivings were coloured by the fact that plays in Athens were performed in the midst of civic rituals rather than in dedicated 'entertainment' venues. Responding to this fear, Plato's student Aristotle influentially prescribed that plays should conclude with a kind of prophylaxis, 'catharsis', which purged the emotions instilled in the audience during the play. But Plato himself preferred a more drastic solution: the right to *poiesis* should be authorised by the philosopher-king. And so were laid the foundations for *censorship* (Jansen 1988).

Here it is worth recalling a distinction drawn by the philologist Erich Auerbach (1953), who taught two of the most influential US literary critics of the late twentieth century, Fredric Jameson and Edward Said. Plato was bothered by what Auerbach called the 'paratactic' character of drama, which calls forth the reader's participation, given the sketchy but suggestive nature of the script. The first moment of participation comes from the actors but once the audience gets involved anything is possible, as the line between fact and fiction becomes blurred. For this reason, dramatists have been historically the most dangerous authors, the social revolutionaries, a point celebrated in Artaud (1958). Auerbach's paradigm case of parataxis was the Bible, which was written to be enacted rather than simply heard – unlike the Homeric epics, which Auerbach dubbed 'hypotactic' for its linguistically self-contained quality. This arguably explains the difference in 'religious' attitudes between the ancient Jews and Greeks, whereby the Jews appeared to believe more strongly in their one deity than the Greeks believed in their many deities. Nevertheless, the scripted character of drama brought the Greeks into the world of the paratactic.

The post-truth condition effectively treats all texts as paratactic. In other words, their truth value is indeterminate prior to their encounter with the reader. On the one hand, the reader may decide that the text is interesting in a self-contained way but ultimately belongs to a world that the reader does not inhabit. The work's existential irrelevance renders it 'fiction'. On the other hand, the reader may choose to inhabit the text by supplying the cognitive and emotional effort required to make it come alive. Such a work is 'fact'. Plato wanted the philosopher-king rather to resolve this interpretive indeterminacy for everyone rather than leave it to individual discretion. In practice, his policy would enforce a clear fact/fiction distinction for each text prior to its encounter with readers.

Here is an example of what Plato meant. Imagine that Darwin's *On the Origin of Species* is officially declared a work of 'fact' rather than 'fiction'. This means that readers are licensed – perhaps even encouraged – to supplement the text's lacunae, vagaries and inadequacies with their own concerns and activities, such as novel observations, discoveries, demonstrations and, of course, authoritative commentary. At the same time, they are discouraged – if not outright prohibited – from using those very same features of Darwin's text as grounds for dismissing it as simply the product of the fevered mind of an author operating with limited capacities at a particular place and time. Moreover, imagine that another work, say, the Bible is declared 'fiction' rather than 'fact'. It is consequently subject to the exact opposite treatment. While the society I have described is not exactly ours, it is reasonably close to it. Indeed, it would be a 'New Atheist' paradise that would prohibit the teaching of religion to children as 'child abuse' for trying to present fiction as fact. And while Plato himself would probably not draw the fact/fiction distinction exactly where – let alone with the same brutality – as the Grand Poobah of the New Atheists, Richard Dawkins does, they would have no trouble recognising each other as kindred spirits.

In contrast, supporters of the post-truth condition would devolve decision-making over what counts as 'fact' and 'fiction' to individuals, who then contest their differences in duly constituted fields of play. In this environment, every text becomes a performative prospect, an invitation to further creative effort, true to the spirit of *poiesis*. A hundred years ago, the original modernist aesthetic movement, Russian Formalism, took this attitude to be the engine of literary innovation (Lemon and Reis 1965). It followed that no work is ever yet complete. There is always still something missing. The story is never quite over. And so one seeks contingency – room to manoeuvre – where others have only been able to see necessity. The ability to imagine such a reversal in modal power led Bismarck to define politics as 'the art of the possible', yet another expression of *poiesis*. This sensibility is also evident in military and economic historians who invoke counterfactuals to show that had certain events shifted slightly, the outcomes would have likely been radically other than they were (Elster 1978: ch. 6; Fuller 1993: ch. 4; Fuller 2015: ch. 6). The implied intent of such imaginative academic endeavours is to mentally prepare the student to take advantage of something similar happening 'for real' in the future.

Indeed, the post-truth approach to history is best seen in the spirit of *method acting* whereby actors draw on their personal histories to convey the relevant emotion in a fully paratactic situation. Such actors need to have lived 'sensitively' but also with an eye to redeploying that experience in a new setting. Thus, professionally trained actors from privileged backgrounds have been able to portray disadvantaged people by finding analogue experiences – however fleeting – in their own past. By virtue of exploiting and embellishing on this emotional wellspring, they become capitalists of emotion whereby *authenticity* becomes the new *sincerity*. By literally 'owning' whatever they have felt in the past, method actors became capable of converting it to capital for future use. According to followers of the great Russian theatre director, Konstantin Stanislavski who invented method acting, this is how to reduce the gap between the audience and the stage – fact and fiction – in one's own person, which in turn is the secret to great acting (Benedetti 1982). It's enough to send Plato back to his allegorical cave!

The post-truth condition projects this attitude on the entire world-historic stage. Not only does it not matter that Dustin Hoffman's father was a Hollywood set designer or that both of Robert De Niro's parents were painters. It also doesn't matter that the proletarian revolution was launched in Russia by a bunch of expatriate intellectuals rather than by the organised labour movement in Germany, as Marx and Engels had originally scripted it in *The Communist Manifesto*. The audience doesn't care, as long as actors bring something vital about themselves into the performance. That is the 'added value' of great acting. *Thus, the performance art known as hypocrisy flourishes because it is more important to own what one says than say what one believes.* By such standards, great politicians are the greatest actors of all, which in turn explains why historians so easily find politicians to have been 'hypocrites'. And many, not least the US founding fathers, had no trouble admitting to hypocrisy, as they easily projected themselves into future scenarios in which others would assume their roles on the world-historic stage (Runciman 2008: ch. 3). More to the point, from the post-truth standpoint, such hypocrisy is not a vice but a virtue.

REFERENCES

Artaud, A. (1958). *The Theatre and Its Double*. New York: Grove Weidenfeld.

Auerbach, E. (1953). *Mimesis: The Representation and Reality in Western Literature*. Princeton: Princeton University Press.

Baudrillard, J. (1983). *Simulations*. New York: Semiotexte.

BBC News (1999). 'Trillions Demanded in Slavery Reparations' (20 August): http://news.bbc.co.uk/1/hi/world/africa/424984.stm

Bell, D. (1960). *The End of Ideology*. New York: Free Press.

——— (1973). *The Coming of Post-Industrial Society*. New York: Basic Books.

Benedetti, J. (1982). *Stanislavski: An Introduction*. New York: Theatre Arts Books.

Berger, P. (1986). *The Capitalist Revolution*. New York: Basic Books.

Berger, P. and Luckmann, T. (1966). *The Social Construction of Reality*. Garden City, NY: Doubleday.

Berlin, I. (1958). *Two Concepts of Liberty*. Oxford: Clarendon Press.

Bernays, E. (1928). *Propaganda*. New York: Horace Liveright.

Bernstein, A. (2019). 'Life, Unlimited: Russian Archives of the Digital and the Human'. *Journal of the Royal Anthropological Institute* 25: 676–97.

Bew, J. (2016). *Realpolitik: A History*. Oxford: Oxford University Press.

Bloom, H. (1973). *The Anxiety of Influence*. Oxford: Oxford University Press.

Bloor, D. (1976). *Knowledge and Social Imagery*. London: Routledge & Kegan Paul.

Boas, F. (1895). *The Social Organization and Secret Societies of the Kwakiutl Indians*. Washington, DC: US National Museum.

Böhm-Bawerk, E. (1898). *Karl Marx and the Close of His System*. London: T. Fisher Unwin.

Bostrom, N. (2014). *Superintelligence*. Oxford: Oxford University Press.

Brandom, R. (1994). *Making It Explicit*. Cambridge, MA: Harvard University Press.

Breitbart, A. (2011). *Righteous Indignation: Excuse Me While I Save the World!* New York: Hatchette.

Briggle, A. (2010). *A Rich Bioethics: Public Policy, Biotechnology and the Kass Council*. South Bend, IN: University of Notre Dame Press.

British Academy (2019). 'The British Academy's President Honours Academics Who Stand Up to Falsehood and Fake News'. https://www.thebritishacademy.ac.uk/news/british-academy-president-honours-academics-stand-against-falsehood-fake-news (18 July).

Butler, J. (1990). *Gender Trouble*. London: Routledge.

Calabresi, G. and Melamed, D. (1972). 'Property Rules, Liability Rules, and Inalienability'. *Harvard Law Review* 85: 1089–128.

Carnap, R. ([1932] 1959). 'The Elimination of Metaphysics through Logical Analysis of Language'. In A. J. Ayer, ed., *Logical Positivism* (pp. 60–81). New York: Basic Books.

Ceci, S. and Peters, D. (2019). 'The Peters and Ceci Study of Journal Publications'. *The Winnower*. doi: 10.15200/WINN.140076.68759

Clark, W. (2006). *Academic Charisma and the Origin of the Research University*. Chicago: University of Chicago Press.

Clifford, W. K. ([1877] 1999). *The Ethics of Belief*. Buffalo, NY: Prometheus Books.

Coates, T-N. (2014). 'The Case for Reparations'. *The Atlantic* (June).

Collins, P. H. (2017). 'Intersectionality and Epistemic Injustice'. In I. Kidd, J. Medina, G. Pohlhaus, eds. *Routledge Handbook of Epistemic Injustice* (pp. 115–24). London: Routledge.

Comfort, N. (2019). 'How Science Has Shifted Our Sense of Identity'. *Nature* 574: 167–70.

Cowen, T. (2006). 'How Far Back Should We Go? Why Restitution Should be Small.' In J. Elster, ed. *Retribution and Reparation in the Transition to Democracy* (pp. 17–32). New York: Cambridge University Press.

Crombie, A. (1996). *Science, Art and Nature in Medieval and Modern Thought*. Cambridge, UK: Cambridge University Press.

Eisenstein, E. (1979). *The Printing Press as an Agent of Change*. 2 vols. Cambridge, UK: Cambridge University Press.

Elster, J. (1978). *Logic and Society*. Chichester, UK: John Wiley.

——— (2004). *Closing the Books: Transitional Justice in Historical Perspective*. Cambridge, UK: Cambridge University Press.

Festinger, L., Riecken, H. and Schachter, S. (1956). *When Prophecy Fails*. Minneapolis: University of Minnesota Press.

Feyerabend, P. (1979). *Science in a Free Society*. London: Verso.

Financial Times (2020). 'Virus Lays Bare the Frailty of the Social Contract'. *Financial Times*, 3 April. https://www.ft.com/content/7eff769a-74dd-11ea-95fe-fcd274e920ca.

Foot, P. (1978). *Virtues and Vices*. Oxford: Oxford University Press.

Forrester, K. (2019). *In the Shadow of Justice*. Princeton: Princeton University Press.

Freud, S. ([1929] 1930). *Civilization and Its Discontents*. London: Jonathan Cape.

Fricker, M. (2007). *Epistemic Injustice*. Oxford: Oxford University Press.

Frye, B. (2016). 'Plagiarism Is Not a Crime'. *Duquesne Law Review* 54: 133–72.

Fuller, S. (1988). *Social Epistemology*. Bloomington, IN: Indiana University Press.

——— ([1989] 1993). *Philosophy of Science and Its Discontents*. 2nd ed. New York: Guilford Press.

———— (2000a). *The Governance of Science*. Milton Keynes, UK: Open University Press.

———— (2000b). *Thomas Kuhn: A Philosophical History for Our Times*. Chicago: University of Chicago Press.

———— (2006a). *The New Sociological Imagination*. London: Sage.

———— (2006b). *The Philosophy of Science and Technology Studies*. London: Routledge.

———— (2007). *New Frontiers in Science and Technology Studies*. Cambridge, UK: Polity.

———— (2009). *The Sociology of Intellectual Life: The Career of the Mind in and around the Academy*. London: Sage.

———— (2010). *Science: The Art of Living*. Durham, UK: Acumen.

———— (2012). *Preparing for Life in Humanity 2.0*. London: Palgrave Macmillan.

———— (2015). *Knowledge: The Philosophical Quest in History*. London: Routledge.

———— (2016). *The Academic Caesar: University Leadership Is Hard*. London: Sage.

———— (2017). 'The Social Construction of Knowledge'. In L. McIntyre and A. Rosenberg, eds. *The Routledge Companion to Philosophy of Social Science* (pp. 351–61). London: Routledge.

———— (2018). *Post-Truth: Knowledge as a Power Game*. London: Anthem.

———— (2019a). 'Against Academic Rentiership: A Radical Critique of the Knowledge Economy', *Postdigital Science and Education* 1: 335–56.

———— (2019b). *Nietzschean Meditations: Untimely Thoughts at the Dawn of the Transhuman Era*. Basel, SZ: Schwabe Verlag.

———— (2020). 'When a Virus Goes Viral: Life with COVID-19'. *Social Epistemology Review and Reply Collective*, 17 March. https://social-epistemology.com/2020/03/17/when-a-virus-goes-viral-life-with-covid-19-steve-fuller/.

Fuller, S. and Collier, J. ([1993, by Fuller] 2004). *Philosophy, Rhetoric and the End of Knowledge*. 2nd ed. Hillsdale, NJ: Lawrence Erlbaum Associates.

Fuller, S. and Lipinska, V. (2014). *The Proactionary Imperative: A Foundation for Transhumanism*. London: Palgrave Macmillan.

Galison, P. (1990). 'Aufbau/Bauhaus: Logical Positivism and Architectural Modernism'. *Critical Inquiry* 16: 709–52.

Gallie, W. B. (1956). 'Essentially Contested Concepts', *Proceedings of the Aristotelian Society* 56: 167–98.

Gambetta, D. (1988). 'Mafia: The Price of Distrust'. In D. Gambetta, ed., *Trust: Making and Breaking Cooperative Relations* (pp. 158–75). Oxford: Blackwell.

Gellner, E. (1959). *Words and Things*. London: Gollancz.

Gilder, G. (1981). *Wealth and Poverty*. New York: Basic Books.

Goldacre, B. (2008). *Bad Science*. London: Fourth Estate.

———— (2012). *Bad Pharma*. London: Harpercollins.

Goldman, A. (1999). *Knowing in a Social World*. Oxford: Oxford University Press.

Goodin, R. and Spiekermann, K. (2018). *An Epistemic Theory of Democracy*. Oxford: Oxford University Press.

Gouldner, A. (1970). *The Coming Crisis in Western Sociology*. New York: Basic Books.

Grafton, A. (1990). *Forgers and Critics: Creativity and Duplicity in Western Scholarship*. Princeton: Princeton University Press.

Gray, J. (2011). *The Immortalization Commission: The Strange Quest to Cheat Death.* London: Penguin.

Guichardaz, R. (2019). 'The Controversy over Intellectual Property in Nineteenth-Century France'. *European Journal of the History of Economic Thought.* December. doi: 10.1080/09672567.2019.1651364

Harari, Y. (2016). *Homo Deus.* London: Harvill Secker.

Harrison, P. (2007). *The Fall of Man and the Foundations of Science.* Cambridge, UK: Cambridge University Press.

Hayek, F. (1945). 'The Use of Knowledge in Society'. *American Economic Review* 35: 519–30.

——— (1952). *The Counter-Revolution of Science.* Chicago: University of Chicago Press.

Holton, G. (1973). *The Thematic Origins of Modern Science.* Cambridge, MA: Harvard University Press.

Innis, H. (1951). *The Bias of Communications.* Toronto: University of Toronto Press.

James, W. ([1896] 1960). *The Will to Believe, Human Immortality and Other Essays in Popular Philosophy* . New York: Dover.

Jansen, S. C. (1988). *Censorship: The Knot That Binds Power and Knowledge.* Oxford: Oxford University Press.

——— (2013). 'Semantic Tyranny: How Edward L. Bernays Stole Walter Lippmann's Mojo and Got Away with It and Why It Still Matters'. *International Journal of Communication* 7: 1094–111.

Kahn, H. (1962). *Thinking About the Unthinkable.* New York: Horizon Press.

Kantrowitz, A. (1977). 'The Science Court Experiment'. *Jurimetrics Journal* 17: 332–41.

Kelly, K. (2014). 'Why You Should Embrace Surveillance, Not Fight It'. *Wired* (10 March).

Kirby, D. (2011) *Lab Coats in Hollywood: Scientists' Impact on Cinema, Cinema's Impact on Science and Technology.* Cambridge, MA: MIT Press.

Koestler, A. (1959). *The Sleepwalkers.* London: Hutchinson.

Koyré, A. (1959). *From the Closed World to the Infinite Universe.* Baltimore: Johns Hopkins University Press.

Kuhn, T. ([1962] 1970). *The Structure of Scientific Revolutions.* 2nd ed. Chicago: University of Chicago Press.

Latour, B. (1987). *Science in Action.* Milton Keynes, UK: Open University Press.

Lemon, L. and Reis, M., eds. (1965). *Russian Formalist Criticism.* Lincoln: University of Nebraska Press.

Lepenies, W. (1988). *Between Literature and Science: The Rise of Sociology.* Cambridge, UK: Cambridge University Press.

Levy, D. and Peart, S. (2017). *Escape from Democracy: The Role of Experts and the Public in Economic Policy.* Cambridge, UK: Cambridge University Press.

Lewis, D. (1969). *Convention.* Cambridge, MA: Harvard University Press.

Lindsay, J., Boghossian, P., Pluckrose, H. (2018). 'Academic Grievance Studies and the Corruption of Scholarship'. *Areo* (2 October) https://areomagazine.com/2018/10/02/academic-grievance-studies-and-the-corruption-of-scholarship/.

Lippmann, W. and Merz, C. (1920). 'A Test of the News'. *The New Republic*. 4 August (pp. 1–42).

Lyotard, J.-F. ([1979] 1983). *The Postmodern Condition* . Minneapolis: University of Minnesota Press.

Maslow, A. (1971). *The Farther Reaches of Human Nature*. London: Penguin.

——— (1998). *Maslow on Management*. New York: John Wiley.

McGoey, L. (2019). *The Unknowers: How Strategic Ignorance Rules the World*. London: Zed.

McLuhan, M. (1964). *Understanding Media: The Extensions of Man*. New York: McGraw Hill.

Meehl, P. (1954). *Clinical vs. Statistical Prediction: A Theoretical Analysis and a Review of the Evidence*. Minneapolis: University of Minnesota Press.

Merton, R. (1973). *The Sociology of Science*. Chicago: University of Chicago Press.

Michels, R. ([1911] 1959). *Political Parties*. New York: Dover Books.

Mirowski, P. and Nik-Khah, E. (2017). *The Knowledge We Have Lost in Information*. Oxford: Oxford University Press.

Nozick, R. (1974). *Anarchy, State and Utopia*. New York: Basic Books.

Oreskes N. and Conway, E. (2011). *Merchants of Doubt: How a Handful of Scientists Obscured the Truth on Issues from Tobacco Smoke to Global Warming*. New York: Bloomsbury.

——— (2014). *The Collapse of Western Civilization: A View from the Future*. New York: Columbia University Press.

Pasquale, F. (2016). *The Black Box Society: The Secret Algorithms That Control Money and Society*. Cambridge, MA: Harvard University Press.

Peters, D. and Ceci, S. (1982). 'Peer-Review Practices of Psychological Journals: The Fate of Published Articles, Submitted Again.' *Behavioral and Brain Sciences* 5 (2):187–95.

Popper, K. (1946). *The Open Society and Its Enemies*. New York: Harper and Row.

——— (1957). *The Poverty of Historicism*. New York: Harper and Row.

Putnam, H. (1979). *Mind, Language and Reality*. Cambridge, UK: Cambridge University Press.

Quine, W. and Ullian, J. (1970). *The Web of Belief*. New York: Random House.

Rawls, J. (1971). *A Theory of Justice*. Cambridge, MA: Harvard University Press.

Rifkin, J. (2011). *The Third Industrial Revolution: How Lateral Power Is Transforming Energy, the Economy, and the World*. New York: Palgrave Macmillan.

Rothblatt, M. (1995). *The Apartheid of Sex*. New York: Crown.

Runciman, D. (2008). *Political Hypocrisy: The Mask of Power from Hobbes to Orwell and Beyond*. Princeton: Princeton University Press.

Sassower, R. (2017). *The Quest for Prosperity: Reframing Political Economy*. London: Rowman and Littlefield.

Scheidel, W. (2017). *The Great Leveller: Violence and the History of Inequality*. Princeton: Princeton University Press.

Schumpeter, J. (1942). *Capitalism, Socialism and Democracy*. New York: Harper & Row.

Sedgwick, E. K. (1990). *Epistemology of the Closet*. Berkeley: University of California Press.

Shapin, S. (1994). *A Social History of Truth*. Chicago: University of Chicago Press.

Soave, R. (2019). 'Portland State University Says Hoax "Grievance Studies" Experiment Violated Research Ethics'. *Reason* (7 January): https://reason.com/2019/01/07/peter-boghossian-portland-irb-hoax-griev.

Sokal, A. and Bricmont, J. (1998). *Fashionable Nonsense: Postmodern Intellectuals' Abuse of Science*. New York: Picador.

Sombart, W. (2001). *Economic Life in the Modern Age*. Eds. N. Stehr and R. Grundmann. New Brunswick, NJ: Transaction Press.

Stark, R. and Bainbridge, W. (1987). *A Theory of Religion*. Berne: Peter Lang.

Tainter, J. (1988). *The Collapse of Complex Societies*. Cambridge, UK: Cambridge University Press.

Taleb, N. N. (2012). *Antifragile*. London: Allen Lane.

Tetlock, P. (2003). 'Thinking the Unthinkable: Sacred Values and Taboo Cognitions'. *Trends in Cognitive Science* 7(7): 320–24.

Tetlock, P. and Gardner, D. (2015). *Superforecasting: The Art and Science of Prediction*. New York: Crown Publishers.

Thompson, G. (2020). *Post-Truth Public Relations*. London: Routledge.

Turner, C. and Behrndt, S. (2007). *Dramaturgy and Performance*. London: Palgrave Macmillan.

Tuvel, R. (2017). 'In Defence of Transracialism'. *Hypatia* 32: 263–78.

Van Norden, B. (2018). 'The Ignorant Do Not Have a Right to an Audience'. *The New York Times* (25 June).

Von Neumann, J. and Morgenstern, O. (1944). *The Theory of Games and Economic Behaviour*. Princeton: Princeton University Press.

Wallerstein, I. (1980). *The Modern World-System II: Mercantilism and the Consolidation of the European World Economy*. New York: Academic Press.

Wark, M. (2004). *A Hacker Manifesto*. Cambridge, MA: Harvard University Press.

Weber, M. ([1918] 1958). 'Science as a Vocation' In H. Gerth and C. W. Mills, eds. *From Max Weber* (pp. 129–58). Oxford: Oxford University Press.

White, H. (1973). *Metahistory: The Historical Imagination in Nineteenth Century Europe*. Baltimore: Johns Hopkins University Press.

Winch, P. (1958). *The Idea of a Social Science*. London: Routledge & Kegan Paul.

Wittes, B. and Chong, J. (2014). *Our Cyborg Future: Law and Policy Implications*. Washington, DC: Brookings Institution.

Wittgenstein, L. (1953). *Philosophical Investigations*. London: Macmillan.

Yates, F. (1966). *The Art of Memory*. London: Routledge and Kegan Paul.

Zahavi, A. and Zahavi, A. (1997). *The Handicap Principle: A Missing Piece of Darwin's Puzzle*. Oxford: Oxford University Press.s

INDEX